THE WANDERING MIND

THE WANDER—ING MIND

WHAT THE BRAIN DOES WHEN YOU'RE NOT LOOKING

MICHAEL C. CORBALLIS

THE UNIVERSITY OF CHICAGO PRESS
CHICAGO

Michael C. Corballis is professor emeritus of psychology at the University of Auckland, New Zealand, and the author of many books, including *A Very Short Tour of the Mind: 21 Short Walks around the Human Brain.*

The University of Chicago Press, Chicago 60637
© 2015 by Michael C. Corballis
All rights reserved. Published 2015.
Printed in the United States of America

24 23 22 21 20 19 18 17 16 15   1 2 3 4 5

ISBN-13: 978-0-226-23861-6 (cloth)
ISBN-13: 978-0-226-23875-3 (e-book)
DOI: 10.7208/chicago/9780226238753.001.0001

LIBRARY OF CONGRESS CATALOGING-IN-PUBLICATION DATA
Corballis, Michael C., author.
  The wandering mind : what the brain does when you're not looking / Michael C. Corballis.
    pages ; cm
  Includes bibliographical references and index.
  ISBN 978-0-226-23861-6 (cloth : alk. paper) — ISBN 978-0-226-23875-3 (e-book) 1. Brain. 2. Cognition. 3. Absent-mindedness. 4. Thought and thinking. I. Title.
  BF201.C67 2015
  153—dc23

                                                      2014035713

A catalogue record for this book is available from the National Library of New Zealand.

Grateful thanks to Alice Duncan-Gardiner for redrawing the hippocampus and seahorse on page 54; and to all other copyright holders for permission to reproduce copyrighted material. Extract from *Speak, Memory* by Vladimir Nabokov © Dmitri Nabokov, 1998, used by permission of the Wylie Agency (UK) Limited.

♾ This paper meets the requirements of ANSI/NISO Z39.48-1992 (Permanence of Paper).

Contents

Preface

Let me see, now, where was I?

Ah, yes, *the wandering mind.*

My wife tells me she was once given the strap at her primary school for looking out the window, probably imagining herself riding her horse. The strap was a form of punishment, now thankfully obsolete, in which a leather strap was brought heavily to bear on the palm of the hand. She claims in mitigation that the boys were also looking out the window, but they remained unpunished. Maybe it was simply accepted that boys were incorrigible mind-wanderers, incapable of paying attention.

Even in adulthood, mind-wandering often elicits feelings of guilt that one somehow didn't pay attention—most commonly, it seems, when introduced to someone by name. Sorry, what was the name again? Perhaps as a result of childhood experiences, many take mind-wandering to be a personal failing. A friend recently asked me to explain why he was unable to concentrate at board meetings, as though he was the only one at fault. I told him it was a sure bet that other minds were wandering just as much. During the day, the evidence shows, our minds are wandering for about half the time. Even at night, when we're asleep, our minds wander into dreams, oblivious to the outside world. Thus, for the sake of us all, but my fellow professors especially, I have a duty to defend absent-mindedness.

To the annoyance of teachers and parents, we seem to be bio-logically disposed to alternate between paying attention and thinking about something else. Mind-wandering, therefore, can't be all bad. Perhaps it's a matter of rest and relaxation, allowing the brain to recover from concentrated activity. Or perhaps it lets some pleasure into our drab lives. But it must surely be more than that. In this book, I argue that mind-wandering has many constructive and adaptive features—indeed, we probably couldn't do without it. It includes mental time travel—the wandering back and forth through time, not only to plan our futures based on past experience, but also to generate a continuous sense of who we are. Mind-wandering allows us to inhabit the minds of others, increasing empathy and social understanding. Through mind-wandering, we invent, tell stories, expand our mental horizons. Mind-wandering underwrites creativity, whether as a Wordsworth wandering lonely as a cloud, or an Einstein imagining himself travelling on a beam of light.

In these pages, I ramble through the various hills and valleys of mind-wandering with the hope of giving it a better name. I have tried to write the nine chapters so that each can be read more or less on its own, although there is a sequence of sorts. I have occasionally allowed myself to wander a bit, but the topic itself seems to permit this. At the same time, I make no claim to have covered all aspects of mind-wandering, nor do I claim to complete accuracy. After all, our minds all wander in different ways.

I have many people to thank. First, Barbara Corballis gave me the first anecdote, and our minds have often wandered together. My sons, Paul and Tim, and my granddaughters, Simone, Lena and Natasha, have contributed in many ways, including giving me more places for my mind to wander to. Among my many professional colleagues I owe thanks to Donna Rose Addis, Michael Arbib, Brian Boyd, Dick Byrne, Suzanne Corkin, Peet Dowrick, Russell Gray, Adam Kendon, Ian Kirk, Chris McManus, Jenni Ogden, Mathias Osvath, David Redish, Giacomo Rizzolatti, Anne Russon and Endel Tulving. A special thanks to Thomas Suddendorf, who provided helpful and constructive comments on a draft of the book.

Finally, thanks to Sam Elworthy and the team at Auckland University Press for their encouragement and faith. I owe a special debt to my editor Mike Wagg, whose fine-tooth comb-over corrected and improved the manuscript in countless ways.

If you're still with me, please read on.

1.
MEANDERING BRAIN, WANDER-ING MIND

• • •

'We're going through!' The Commander's voice was like thin ice breaking. He wore his full-dress uniform, with the heavily braided white cap pulled down rakishly over one cold gray eye. 'We can't make it, sir. It's spoiling for a hurricane, if you ask me.' 'I'm not asking you, Lieutenant Berg,' said the Commander. 'Throw on the power lights! Rev her up to 8,500! We're going through!' The pounding of the cylinders increased: ta-pocketa-pocketa-pocketa-*pocketa-pocketa*. The Commander stared at the ice forming on the pilot window. He walked over and twisted a row of complicated dials. 'Switch on No. 8 auxiliary!' he shouted. 'Switch on No. 8 auxiliary!' repeated Lieutenant Berg. 'Full strength in No. 3 turret!' shouted the Commander. 'Full strength in No. 3 turret!' The crew, bending to their various tasks in the huge, hurtling eight-engined Navy hydroplane, looked at each other and grinned. 'The old man will get us through,' they said to one another. 'The Old Man ain't afraid of Hell!' ...

'Not so fast! You're driving too fast!' said Mrs. Mitty. 'What are you driving so fast for?'

So begins James Thurber's short story 'The Secret Life of Walter Mitty', first published in *The New Yorker* in 1939, and made into a film starring Danny Kaye in 1947 (a recent remake stars Ben Stiller). Mitty is the archetypal daydreamer. He is, of course, a fictional character, so he, and his mind-wandering, are really products of Thurber's own mind-wandering. The wandering mind is the source of fiction, as well as possible mishaps on the road.

The Chambers dictionary defines wandering in several ways, but the one I like best goes like this:

Wander /won'der/ intransitive verb. To go astray, deviate from the right path or course, the subject of discussion, the object of attention, etc.[*]

That definition seems to allow that we can wander mentally as well as physically. Mind-wandering often seems to afflict us when we're supposed to be concentrating on something, such as a lecture, a board meeting, or driving. It also gets in the way when we're simply trying to read a book. Jonathan Schooler and colleagues at the University of California at Santa Barbara had students read the opening chapters of Tolstoy's *War and Peace* for 45 minutes and asked them to press a key whenever they caught themselves 'zoning out'. They caught themselves an average of 5.4 times. The students were also interrupted six times at random intervals to see if they were zoning out at the time without having been aware of it, and this caught, on average, a further 1.2 times. So it's not just you, you might be relieved to know—we all seem to have trouble staying focused, especially on the books we're actually supposed to be reading. Or the lecturer we're supposed to be listening to.

Okay, you can now zone back in.

Sometimes, mind-wandering is intrusive even when you're not engaged in some more pressing task. Perhaps you've been on a long plane trip, trying to sleep. Somehow the mind won't turn off, but churns through tedious or worrisome thoughts. You might brood over some recent unsettling incident, or fret about a forthcoming

---

[*] In German, I'm told, 'wandern' simply means 'to walk', without any suggestion of deviating from the right path. Germany has a fine literary and artistic tradition, indicating that German minds can also wander in the sense intended in this book.

lecture. Of course, our wanderings can also be happy—anticipating a family reunion, or luxuriating in a recent promotion. Sometimes, too, the mind gets caught in an eddy, with thoughts that keep repeating.

As often as not, the eddy is a tune or jingle that won't go away. It's like a stuck record. This has been called 'stuck song syndrome', and the offending jingles are known as 'earworms'. The problem is how to get rid of them. One suggestion is to dump them on someone else. In his short story 'A Literary Nightmare', published in 1876, Mark Twain tells of a virus-like jingle that infects his mind for several days, until he goes for a walk with his friend the Reverend, and manages to transfer it to him. Twain later meets up with the Reverend again and finds him in distress—the jingle has so infected his thoughts and actions that the congregation in his church has started swaying to its rhythm. Twain takes pity on him and helps him transfer the jingle to a group of university students.

The jingle in question was based on a sign announcing the fares on a tram, and converted into a short song with a catchy tune. It runs as follows (skip this if you don't want to be infected):

Conductor, when you receive a fare,
Punch in the presence of the passenjare!
A blue trip slip for an eight-cent fare,
A buff trip slip for a six-cent fare,
A pink trip slip for a three-cent fare,
Punch in the presence of the passenjare!
CHORUS
Punch brothers! Punch with care!
Punch in the presence of the passenjare!

The jingle seems to have gone on to infect popular culture, at first in Boston and especially among Harvard students, and then more widely. It was even translated into French and Latin. It was used by Robert McCloskey in one of his Homer Price stories, called 'Pie and Punch and You-Know-Whats', as a cure for another persistent jingle. It was set to music in 1972 as part of a song cycle called *Third Rail* by Donald Sosin, and performed at several venues in the United States. The jingle has no doubt faded from the public mind as other annoying jingles have risen to take its place, which are best not mentioned here in case they stick and you can't get rid of them.

What the brain does while the mind wanders

The brain is active even when the mind is disengaged, or wandering from the task at hand. Early evidence that this is so came as a result of a German physician called Hans Berger (1873–1941) falling from his steed and landing in the path of a horse-drawn cannon. Fortunately for the future of neuroscience, he escaped injury, but his sister at home several kilometres away sensed that he was in danger, and asked her father to contact him. Berger took this as evidence for telepathy, which he thought might depend on some physical transmission of 'psychic energy', and might be measurable. In 1924, he decided to test this by recording electrical activity from two electrodes placed under the scalp, one at the front of the head and one at the back. Sure enough, the electrodes picked up electrical activity, although it was far too weak to suggest a basis for telepathy. The technique became known as electroencephalography (EEG). When the subject was in a resting state with the eyes closed, the EEG showed a fluctuation in voltage with a frequency of 8 to 13 cycles per second, known then as 'Berger's wave' but more recently as the 'alpha wave'. When the eyes were opened, this wave was suppressed by the faster 'beta wave'. In later developments of electroencephalography, multiple electrodes are placed on the surface of the scalp, and can provide information as to where in the brain the activity is generated.

Later on, better techniques for witnessing activity in the brain were invented. In the 1970s, the Swedish physiologist David H. Ingvar, along with Danish scientist Niels A. Lassen, injected a radioactive substance into the bloodstream and tracked its course in the brain with external monitors. The blood flows to regions where neural activity is high, and Ingvar noted that activity was especially high in the frontal areas of the brain during resting states. He described this as representing 'undirected, spontaneous, conscious mentation'. In short, mind-wandering.

Since then, increasingly sophisticated methods of tracking blood flow and superimposing it on detailed anatomical images of the brain have given us much more precise maps. One technique, known as positron emission tomography (PET), also uses injection of radioactive substances in the bloodstream, while a less invasive technique known as functional magnetic resonance imaging (fMRI) uses a powerful magnet to detect haemoglobin, which is carried by the blood. In both cases, the movement of blood is super-imposed on images of the structure of the brain. These techniques are used in clinical research to investigate brain pathology, but fMRI in particular is increasingly used in normal volunteers to map the brain networks involved in simple mental tasks such as reading, or recognising faces, or mentally rotating objects.

These techniques have allowed us to see which parts of the brain are active, both when a person is engaged in a task and when she is not. At first, it was assumed that activity during non-engagement was simply background neural noise, like static on an old radio. In studying the brain activity associated with a given task, such as reading words, it was supposed that one could simply subtract out the neural signal when the brain was idling from that when it was engaged in the task. It transpired, though, that blood flow to the idling brain was only 5 to 10 per cent lower than to the engaged brain, and wider regions of the brain were active during idling than during engagement on a task. The brain regions active during the supposedly resting state have come to be known as the 'default-mode network'. It was Marcus Raichle, a neurologist from Washington University in St Louis, Missouri, who coined this term, in 2001. 'To my great surprise,' he wrote to me, 'it has taken on a life of its own, for better or worse.'

The default-mode network covers large regions of the brain, mainly in the areas not directly involved in perceiving the world or responding to it. The brain is a bit like a small town, with people milling around, going about their business. When some big event occurs, such as a football game, the people then flock to the football ground, while the rest of the town grows quiet. A few people come from outside, slightly increasing the population. But it's not the football game we're interested in here. Rather, it's the varied business of the town, the give and take of commerce, the sometimes meandering activity of people in their communities and places of work. So it is in the brain. When the mind is not focused on some event, it wanders.

Mind-wandering can be under conscious control, as when we

deliberately replay past memories or plan possible future activities. Sometimes it is involuntary, as when we dream, or hallucinate—things that just happen to us whether we want them to or not. Sometimes it lies somewhere in between, as when we daydream, perhaps with the intent of considering some dilemma, or try to solve a cryptic crossword clue, but other thoughts intrude. As the American comedian Steven Wright complained: 'I was trying to daydream but my mind kept wandering.'

Mind-wandering plays cat and mouse with paying attention. In one study, Japanese researchers had people watch videos while recording their brain activity. Most of the time, the brain areas concerned with paying attention were active, but at natural breaks in the stream of events people would blink their eyes, and brain activity would momentarily shift to the default-mode network. Indeed, sometimes when people are supposed to be paying atten-tion to something like a video, they blink more often than is necessary to lubricate the eyes. This is a sign that their minds have flitted away from the story.

Is mind-wandering bad for you?

Some have suggested that mind-wandering is not good for us, and one study suggests that it even makes us unhappy. The authors of this study exploited the age of the smartphone by devising an app that enabled them to contact around 5000 people from 83 different countries at random moments through their waking hours, and ask them what they were doing when interrupted. In 46.7 per cent of the samples, they were thinking about something other than what they were currently doing—or supposed to be doing. Their minds, in other words, were wandering. They were more likely to have wandered into pleasant thoughts than into unpleasant ones, and not surprisingly reported being happier when basking in the pleasantness than when grovelling in the unhappiness. But even when wandering into pleasant thoughts they were no happier than when their minds were not wandering. The study's authors assert: 'A wandering mind is an unhappy mind.' Perhaps, though, their happiness was diminished by being so rudely interrupted.

In one respect, at least, it is not surprising that we can be happier when actually engaged in an activity than when simply imagining it. In the study just mentioned, the activity that created the greatest happiness was making love. Simply imagining that happy activity understandably fails to match the bliss of the real thing. Well, most of the time, anyway. More generally, we may concoct joyful plans, but the joy is capped by the satisfaction of carrying them out. Conversely, events we dread often turn out to be less catastrophic than we feared.

Even so, there's more bad news. People whose minds wander a lot seem to have shorter telomeres (the repeated nucleotides at the ends of chromosomes) in immune cells, which is taken to be a sign of aging. Too much worry and introspection, and you'll die young. If I were you, though, I'd be inclined to take that with a grain of salt—although do remember that salt, too, increases the risk of cardiovascular disease and early death.

So you may well wonder why nature equipped us with wandering minds. Besides possibly creating unhappiness and premature death, our mental journeys are a hazard to driving and a general impediment to efficient performance. They may lead us to fail our exams, forget our appointments, leave the stove on as we depart for a vacation. Teachers implore their pupils to pay attention, to stop dreaming, so that they might learn, and part of the unhappiness caused by mind-wandering may stem from the guilt at having been reprimanded for inattention in our youth.

As adults, we feel guilty that our minds are not on the job, perhaps when we're supposed to be marking a heap of exam scripts or sorting envelopes. Most people, it seems, find their jobs boring at times, and wish they were somewhere else, but then feel guilty for doing so. The rather bad press associated with mind-wandering seems to have stirred an interest in what has been called 'mindfulness'—a form of meditation in which we turn our thoughts intensely inward and remain locked in the present. The Buddha is said to have advised as follows:

> The secret of health for both mind and body is not to mourn for the past, worry about the future, or anticipate troubles, but to live in the present moment wisely and earnestly.*

Rather than allow our minds to roam around the mental landscapes of past and present, or gardens of joy or anguish, we are urged to remain within our own skins, moving a spotlight of attention from one part of the body to another, or intently examining the sensations of breathing. I have no doubt that such techniques can restore a mental calm, although one may well wonder whether

---

* Actually, this quote seems to come from the translation of a Japanese book called *The Teaching of Buddha*, and placed in hotel rooms as a Buddhist alternative to the Gideon Bible. Don't worry too much about it.

mindfulness, any more than mind-wandering, actually helps us get focused on the things we must do.

In any event, the news about mind-wandering is not all bad. Italian researchers found that excessive mind-wandering, even when shorn of what they call 'perseverative cognition'—rumination and worry—may have negative effects on health in the short term, but no detectable effects a year later. It seems we are programmed to alternate between mind-wandering and paying attention, and our minds are designed to wander whether we like it or not. In adapting to a complex world, we need to escape the here and now, and consider possible futures, mull over past mistakes, understand how other people's minds work. Above all, mind-wandering is the source of creativity, the spark of innovation that leads in the longer run to an increase rather than a decrease in well-being. It is even suggested that we have entered a new era of education that recognises creativity and problem-solving, rather than simply 'drilling the rote memorisation of facts and figures'. Maybe we should stop feeling guilty about mind-wandering and learn to revel in our Mitty-ish escapades.

In the following chapters, I discuss some of the components of mind-wandering, often with an eye to its likely adaptiveness and evolutionary origins. I will suggest that even rats may indulge in mental perambulations. But I begin with the faculty that must lie at the heart of our wandering minds. It's called memory.

2.
MEMORY

To fix correctly, in terms of time, some of my childhood recollections, I have to go by comets and eclipses, as historians do when they tackle the fragments of a saga. But in other cases there is no dearth of data. I see myself, for instance, clambering over wet black rocks at the seaside while Miss Norcott, a languid and melancholy governess, who thinks I am following her, strolls away along the curved beach with Sergey, my younger brother. I am wearing a toy bracelet. As I crawl over those rocks, I keep repeating, in a kind of zestful, copious, and deeply gratifying incantation, the English word 'childhood', which sounds mysterious and new, and becomes stranger and stranger as it gets mixed up in my small, overstocked, hectic mind, with Robin Hood and Little Red Riding Hood, and the brown hoods of old hunchbacked fairies. There are dimples in the rocks, full of tepid seawater, and my magic muttering accompanies certain spells I am weaving over the tiny sapphire pools.

The place is of course Abbazia, on the Adriatic. The thing around my wrist, looking like a fancy napkin ring, made of semitranslucent, pale-green and pink, celluloidish stuff, is the fruit of a Christmas tree, which Onya, a pretty cousin, my coeval, gave me in St. Petersburg a few months before. I sentimentally treasured it until it developed dark streaks inside which I decided as in a dream were my hair cuttings which somehow had got into the shiny substance together with my tears during a dreadful visit to a hated hairdresser in nearby Fiume.

—Vladimir Nabokov, from *Speak, Memory*

In one way or another, all of our mind-wanderings depend on memory. Without memory, we would have nowhere for our minds to wander to. It provides the material that feeds the imagination, allowing us to visit the past, and construct futures and fantasies. Even the chaotic things that happen in dreams depend on people, places, events, triumphs and tragedies recorded from the past, often combined and blended in random or bizarre ways. To begin our foray into mind-wandering, then, we need to consider how memory itself works.

Memory is not simple, and consists of at least three layers. At the most basic level, are the skills that we learn. We learn to walk, talk, write, ride a bike, play the piano, play tennis, tap out messages on our smartphones. In varying degrees, we are genetically disposed to do these things. Given normal physical endowment, the ability to walk unfolds in childhood largely automatically, but the infant nevertheless spends time practising and perfecting her new-found capacity for getting around. The ability to talk, too, is an innately human capacity, but the actual languages we acquire, and even the particular sounds we make as we speak, depend on experience. Some 7000 languages exist in the world, made up of rather different sound patterns, and we are all locked in to just one or two of these linguistic stockades. Even languages that seem alike tend to drift apart;* and parents find their children increasingly difficult to understand as they venture into adolescence.

Once learned, skills tend to stay with us. It's said that you never forget how to ride a bike, although old age and arthritis eventually take their toll. Yet some skills, especially those learned late, can be lost. I once accompanied my four-year-old son to recorder lessons

---

* George Bernard Shaw once remarked: 'England and America are two countries separated by the same language.'

and learned to play rather badly, but I now find I can't remember a single configuration. Even language tends to fray, and words become more elusive as we dip into old age. Other skills become unlearnable. In early childhood, we learn any language with ease, but as adults we struggle to acquire any foreign language, especially if its sound patterns are distant from our own. I watch teenagers texting on their smartphones, thumbs flitting across the tiny keyboards, and know that I will never acquire that degree of skill.

We take our skills into our mind-wandering, sometimes with a degree of expertise now lost. I occasionally dream of playing squash or field hockey with the (modest) proficiency of old, but physically they'd now leave me for dead—perhaps literally. Watching a rugby game, I can imagine myself edging through a gap to score, or placing a strategic kick, but these are now just fantasies. One of the delights of mind-wandering is that it allows us to recover former skills. But perhaps it's also a source of the unhappiness, mentioned in the previous chapter, that is recorded when we are abruptly forced back into the present, and the dream is snatched away.

The next level of memory is knowledge, our storehouse of facts about the world. Our knowledge acts as a sort of combined encyclopaedia and dictionary, and is a huge storage system. For a start, it contains all the words we know, as well as what they stand for. Readers of this book probably have a vocabulary of at least 50,000, and that's also a fair count of the number of objects, people, actions, qualities and so forth that we know and talk about. We know places—cities, beaches, ski slopes, favourite cafés. We know facts we learned in school—third-declension Latin nouns, the boiling point of water, how photosynthesis works. People write whole books on what they know, as I am struggling to do now.

We also know lots of things about the people we know, such as what they do, where they live, what their habits are, what kind of tennis game they have, whether they cheat at cards. We even know a bit about ourselves, perhaps embellished and not always consistent with what others know about us. The poet Edward Lear may or may not have been accurate when he wrote of himself:

How pleasant to know Mr. Lear,
Who has written such volumes of stuff.
Some think him ill-tempered and queer,
But a few think him pleasant enough.

His mind is concrete and fastidious,
His nose is remarkably big;
His visage is more or less hideous,
His beard it resembles a wig.

And so it goes on, as a mixture, no doubt, of fact and mind-wandering.

Most of our knowledge is long-lasting and stable, although there are lots of facts we forget. Much of our knowledge comes from schooling, but how much of what you learned in high school or university do you remember? Not much, you may think. Some of it trickles back, though, when your children go through schooling and seek your help, and you recover long-unused knowledge, like Newton's laws of motion or the dates of the French Revolution. But even though much of our early learning seems to have evaporated, our knowledge is vast, and surely a hallmark of being human. The Greek poet Archilochus (c. 680–c. 645 BC) is credited with the saying 'The fox knows many things but the hedgehog knows one big thing', but we humans eclipse them both—or at least we think we do.

The third layer is memory for the specific events of our lives, commonly known as episodic memory. It is in this sense that we generally use the term *remember*, and remembering is itself part of mind-wandering. Unlike knowledge, which is essentially a static system that provides us with information, remembering is a dynamic re-enactment of the past. Given that the things we remember are essentially personal, they may be said to make up much of what we understand as the self. The things we know are for the most part shared with others, but our episodic memories make each of us unique.

Although we do lose some of our skills, and once-held pieces of knowledge occasionally elude us, it is episodic memory that is the most fragile of the three stages. Such is the length and complexity of our conscious lives that we probably retain only a tiny fraction of the things that happened to us. The émigré Czech writer Milan Kundera, in his novel *Ignorance*, put it as follows:

> The fundamental given is the ratio between the amount of time in the lived life and the amount of time from that life that is stored in memory. No one has ever tried to calculate this ratio, and in fact there exists no technique for doing so; yet without much risk of error I could assume that the memory retains no more than a millionth, a hundred-millionth, in short an utterly infinitesimal bit of the lived life. That fact too is part of the essence of man. If someone could retain in his memory everything he had experienced, if he could at any time call up any fragment of his past, he would be nothing like human beings:

neither his loves nor his friendships nor his angers nor his capacity to forgive or avenge would resemble ours.

Well, he does exaggerate a bit. A hundred-millionth would amount to only about 15 minutes of remembered events, and most of us can do much better than that.

We can, if pressed, produce quite a large number of episodes from the past. In the laboratory run by my colleague Donna Rose Addis, we borrow an idea from the board game Cluedo, in which contestants vie with each other to discover who committed a murder, with what instrument, and where—it might have been the Reverend Green, with a candlestick, in the billiard room. We ask our subjects to recall around 100 episodes from the past, and identify a person, an object and a location involved in each episode. (We then scramble these components and ask the subjects to generate new episodes, but I'll talk more about that in the next chapter.) Our subjects have little difficulty recalling the required number of events. When we put our minds to it, we can remember lots of events from the past, and even write autobiographies covering the best part of a lifetime. But most of us are unaware of the vast number of events that we have forgotten, precisely because we've forgotten them! Thumbing through old photograph albums can reveal scenes that seem to belong to another life.

Amnesia

The fragility of episodic memory is well illustrated by cases of amnesia, where it is memory for past events that is typically most affected, and in some cases completely wiped out. The most extensively studied case was a man long known in the literature as 'H.M.', but whose real name was more recently revealed to be Henry Molaison. Henry was probably the most famous case in the history of neurology, and when he died in 2008 at the age of 82, obituaries to him were published in the *New York Times* and in the respected medical journal *The Lancet*. At the age of 27 he underwent surgery for intractable epilepsy, and it was the surgery that was mostly to blame for destroying the parts of his brain responsible for recording personal memories. He formed no memories back in time to the operation, and remembered little of his earlier life as well. He remained able to talk normally, and his IQ was above normal. The authors of a report written in 1968 write: 'His comprehension of language is undisturbed: he can repeat and transform sentences with complex syntax, and he gets the point of jokes, including those turning on semantic ambiguity.'

Over the years since that fateful operation, Suzanne Corkin, first as a graduate student at McGill University and later as a professor at MIT, carried out most of the testing on Henry and came to know him well. He never got to know her, though, and always greeted her anew, recounting the same few stories that he could remember from his childhood. With curious insight, he once described his relationship with Corkin: 'It's a funny thing—you just live and learn. I'm living and you're learning.' A flavour of his character and condition also comes from the following snippet of conversation with my one-time colleague Jenni Ogden:

Jenni: How old do you think you are now?
Henry: Round about 34. I think of that right off.

Jenni: How old do you think I am?

Henry: Well, I'm thinking of 27 right off.

Jenni: (*laughing*) Aren't you kind! I'm really 37.

Henry: 37? So I must be more than that.

Jenni: Why? Do you think you're older than me?

Henry: Yeah.

Jenni: How old do you think you are?

Henry: Well, I always think too far ahead in a way. Well, nearer, well, 38.

Jenni: Thirty-eight? You act 38! You know, you are really 60. You had your 60th birthday the other day. You had a big cake.

Henry: See, I don't remember.

Rather surprisingly, Henry could draw an accurate representation of the floor plan of the house he moved into after the operation, although it took many years for this memory to be established. He therefore retained some ability to acquire new knowledge. He was also able to learn new skills, even though unable to remember the learning episodes themselves. One example is mirror-tracing. He was asked to trace a five-pointed star, keeping the pencil inside its boundaries, but could only see the star and his hand in a mirror. This is quite difficult (try it), since the movements you need to make are front–back reversed from what you see in the mirror. Henry improved rapidly over successive days. On the last day, having easily traced the star, he said: 'Well, this is strange. I thought that it would be difficult, but it seems as though I've done it quite well.'

Another striking case is the English musician Clive Wearing, an expert in early music who built up a distinguished career with the BBC. He founded the Europa Singers, an amateur choir that went on to achieve considerable success, and was responsible for the musical content on Radio 3 on the day of the royal wedding of Prince Charles to Diana Spencer. In 1985, at the height of his career, he was struck down by herpesviral encephalitis, caused by a form of *Herpes simplex* (the cold-sore virus) that very occasionally attacks the central nervous system. It was some time before the condition was diagnosed and he was given drugs to halt the damage, but by this time the areas critical to the formation of new memories, along with some of his already established memories, were eradicated.

At least some of his prior skills and knowledge were retained. He could still talk, play the piano, and conduct a choir. He knew he was married, but could not recall the wedding; he knew he was a musician, but could not recall any concert. Nevertheless, large chunks of his prior knowledge, especially that relatively close to his illness, were gone. He recognised his children but expected them to be much smaller, and wasn't sure how many he had. He didn't know the year or the decade, and was surprised to see that *The Times* no longer had personal columns on the front page. He knew facts about his childhood, where he grew up, where he was evacuated to during the war. He even knew he went to Clare College, Cambridge, on a scholarship. But he was unable to acquire new knowledge, and his storehouse of knowledge was thrust back years in time.

It was his episodic memory, though, that was completely snatched from him. He lives in a window of a few seconds, enough to sustain something of a conversation, although he soon forgets topics he spoke about only moments earlier. The title of a 2005 ITV documentary in England described him as 'The Man with the 7 Second Memory'. Indeed, his window of memory is so short

that he is often surprised at things he has just done. He likes to play patience, and having laid out cards and then shuffled the ones in his hand, he would be startled to see the cards that were laid out. 'And the cards,' he would say, 'they're not laid out by me! I've never seen them before! I can't understand it . . . The world's gone mad!'

Yet another well-known case, identified in the literature simply as 'K.C.', has little difficulty with factual knowledge, but can't remember particular events from his past. It's not just that events are one-off happenings that are not rehearsed. K.C. couldn't remember events that lasted several days, such as being evacuated from home, along with tens of thousands of others, when a derailment nearby released toxic chemicals. Otherwise he scores normally on intelligence tests and knows the basic facts of his life, many of them seldom rehearsed. He knows his date of birth, the address of the home he lived in for the first nine years of his life, the names of schools he went to, the make and colour of the car he once owned, the location of a summer cottage his parents own and its distance from his home in Toronto. He knows lots, but remembers little.

A condition that has much the same effects on memory is Korsakoff syndrome, one of the consequences of chronic alcoholism. In his book *The Man Who Mistook His Wife for a Hat*, Oliver Sacks writes of a case called Jimmie G., who could remember nothing since the end of World War II, and even in the early 1980s believed it was still 1945. He never gets over the shock of seeing his face in the mirror, since he expects to see a young, fit man in his twenties. One of the few benefits of drinking too much, then, is that you can believe you are much younger than you are, so long as you keep away from mirrors.

In one respect, at least, these cases of amnesia are limited in their ability to mind-wander—they are denied access to the past. They have lost the luxury of nostalgia.

Super-memory

If we sometimes despair of having a poor memory, it would also be a serious impediment to remember everything. Our minds would be so clogged as to leave little room for much else. A condition known as 'savant syndrome' can result in prodigious powers of memory, but deficiencies in other aspects of intelligence. One extraordinary example is Kim Peek, the inspiration for the movie *Rain Man*. He died, aged 58, in 2009. Known to his friends as 'Kim-puter', he began memorising books at the age of 18 months. By his mid-fifties he had memorised 9000 books. He had a vast storehouse of knowledge in history, sports, movies, space programmes, literature and Shakespeare, among other things. He had an extensive knowledge of classical music, and in middle life had even begun to play it. Like other savants, he could tell you at once the day of the week for any given date, a feat known to depend on massive memory.

Yet on a standard test of intelligence, Peek scored only 87 (the population average is 100). He had an unusual sidelong gait, could not button his clothes, and could not handle the chores of daily life. He also had difficulty with abstract ideas. What this profile suggests is that a large and detailed memory can work to the disadvantage of other mental skills, and a memory that is too particular can impair ability to see relations and form abstractions. Too many trees, and it's hard to see the wood.

Another remarkable savant is Daniel Tammet, famous for having learned to speak Icelandic in a week, a feat he accomplished for a TV documentary. In March 2004, he recited the mathematical constant *pi* (the ratio of a circle's circumference to its diameter) to 22,514 decimal places. For this, he depended on seeing numbers in his mind's eye as 'complex, multidimensional, coloured, and textured shapes'. This ability to associate entities in one sensory domain with qualities in another is known as 'synaesthesia'. Tammet, then, was able to see the digits of *pi* roll by as a numerical panorama, 'the beauty of which both fascinated and enchanted me'. He also found poetry in his synaesthetic vision. A verse from a poem he wrote based on a visit to Iceland goes like this:

And in the towns and cities
I watched people talking among themselves
Stitching their breath
With soft and coloured words.

A rather different case was Solomon Shereshevskii, also known in the literature as 'S', whose prodigious feats of memory were described by the great Russian neuropsychologist Aleksandr Romanov Luria in his 1968 book *The Mind of a Mnemonist*. Shereshevskii's memory capacity was seemingly without limit, and he remembered trivial things for extremely long periods of time. He could accurately recall lists of words that Luria had presented sixteen years earlier. His memory was mainly visual, and when given words or numbers to remember he could transform them mentally, either by arranging them in spatial patterns, or using the 'method of loci' whereby he would imagine them in familiar locations and later 'play them back' by visiting those locations in his mind.

The particularity of his memory was actually an impediment, because it prevented him from forming general concepts. He couldn't make sense of novels, since he would imagine scenes in precise detail, only to find his images contradicted at later stages. Like the Russian novelist Vladimir Nabokov, who opened this chapter, and Daniel Tammet, he too was a synaesthete, so that spoken words were accompanied by visual sensations, such as 'puffs' or 'splashes', and a tone at precisely 30 cycles per second and 100 decibels gave rise to 'a strip 12–15 cm in width the color of old tarnished silver'.

He should be so lucky, you might think—not all of us can conjure images of old tarnished silver. But in fact his extraordinary memory and intrusive visual imagery were serious impediments to a normal life. Luria quotes an example related to him by Shereshevskii:

> One time I went to buy some ice cream … I walked over to the vendor and asked her what kind of ice cream she had. 'Fruit ice cream,' she said. But she answered in such a tone that a whole pile of coals, of black cinders, came bursting out of her mouth, and I couldn't bring myself to buy any ice cream after she had answered in that way.

The method of loci, though, need not be allied with synaesthesia, and is a useful technique that anyone can learn, although probably not to anything like the degree achieved by Shereshevskii. It is in fact a practical application of mind-wandering, albeit in a controlled form. According to Cicero, it was discovered by a Greek poet named Simonides, who was entertaining a group of wealthy noblemen at a banquet, when he was called outside by two mysterious figures, who turned out to be messengers from the Olympian gods Castor and Pollux. As soon as he left, the roof of the banquet hall collapsed, killing all those inside. The bodies were too mangled to be identified, until Simonides came forward and indicated where each had been sitting, and was then able to name each in turn. On the basis of this story, Greek and Roman orators were said to use the method of loci to memorise their speeches.

The method of loci was adapted by Matteo Ricci, an Italian Jesuit missionary working in China. In 1596, he wrote a book called *Treatise on Mnemonic Arts*, setting out a technique to enable Chinese men to retain the vast knowledge they needed to pass the civil service examinations. It was based on an imagined 'memory palace', made up of a reception hall and many rooms with vivid images, depicting such emotional scenes as war or religious events. The idea was to associate items to be remembered with these images, often in emotional or outrageous combinations, so that later a mental wander through the palace would reveal what was to be remembered.

Even today, the method of loci seems to be the mnemonic of choice for the world's top memorisers. One who uses it is Lu Chao, a Chinese businessman who holds the Guinness record for reciting the decimal places of the constant *pi*. In 2006, he recited *pi* to 67,890 decimal places, before making an error on the 67,891st, tripling the earlier record set by Tammet. An even more striking example, though, is a young engineering student, perhaps not surprisingly (but maybe jokingly) referred to in the literature as 'PI'. He recited *pi* to more than 2^{16} decimal places. He is said to have made 'under 2^4 errors', which may sound a lot, but on average is only one error per 2^{12} digits. Just why he made occasional errors is not stated, but may have to do with some fuzziness in the imagined locations. Shereshevskii occasionally failed to remember an item because he had mentally placed it in a rather dark location, but this could sometimes be corrected when he imagined turning a street light on.

These last cases, unlike Shereshevskii, seem otherwise normal, although PI has a rather poor memory for events and for neutral faces—he seems better on faces showing emotion. Such cases aside, it may be that techniques like the method of loci have become largely irrelevant. You can download *pi* to whatever precision you need from your iPad. And who needs it to 2^{16} decimal places?

False memories

> I have been through some terrible things in my life, some of which actually happened.
> —Mark Twain[*]

Our memories are not only incomplete, they are also often inaccurate, and we sometimes 'remember' things that didn't actually happen. The American psychologist Elizabeth Loftus, a pioneer in the study of false memories, vividly recalls her own mother's death. She was fourteen years old, and visiting her aunt and uncle. She remembers the fateful day as bright and sunny, and recalls the sight and smell of cool pine trees, the taste of iced tea. She sees

---

[*]   The remark is widely attributed to Mark Twain, but he probably didn't make it. It may itself be a false memory.

her mother in her nightgown, floating face down, drowned in a swimming pool. She cries out in terror, screams, sees police cars with lights flashing, and watches her mother's body being carried out on a stretcher. But the memory is false. She was in fact asleep when the body was found, not by her but by her aunt Pearl.

I have a strong memory of a famous rugby match in 1981 when New Zealand played South Africa in Auckland, amid protests against the apartheid regime in South Africa. Two demonstrators in a light plane dropped flour bombs on the players, one of which struck and felled the All Black Murray Mexted—an event that I thought might explain the malapropisms that found their way into Mexted's vocabulary when he later became a rugby commentator. Alas for my theory, and for my memory, I later discovered that it was not Mexted whom the flour bomb hit. It was Gary Knight, whose vocabulary, as far as I know, is fine.

False memories are easily implanted. People often give detailed answers when asked to describe being lost in a mall, or being taken for a ride in a hot-air balloon, or being nearly drowned and rescued by a lifeguard, even though these events never actually happened to them. In another example described by Loftus, people were shown a fake ad extolling a visit to Disneyland which included mention of Bugs Bunny. About a third of them claimed they had themselves visited Disneyland and shaken hands with Bugs Bunny. They could see it in their mind's eye. Bugs Bunny, though, is the creation of Warner Brothers and does not feature in the property of The Walt Disney Co. The memory was false.

Jean Piaget, the famous Swiss psychologist, clearly recalled an incident when he was four years old. His nurse was pushing him in a pram along the Champs Élysées in Paris, when a man tried to kidnap him. He was strapped in, and the nurse tried to stand between him and the kidnapper. In the scuffle, she was scratched on her face, and Piaget claimed he could still see the scratches in his mind's eye. When he was fifteen, though, his nurse wrote to him to say she had made up the whole story.

In the late nineteenth century, false memories were known as 'paramnesias', and it was known that they could be induced through hypnosis. An especially horrific example is described by the hypnotherapist Hippolyte Bernheim, who tells how he suggested to a patient that she had watched through a keyhole an old man raping a little girl, who struggled, was bleeding, and was then gagged. His suggestion ended: 'When you wake up you will think no more about it. I have not told the story to you; it is not a dream; it is not a vision I have given you during your hypnotic sleep; it is truth itself.' Three days later, Bernheim asked a distinguished lawyer friend to question the patient. The patient repeated the events in detail as suggested to her, and even when encouraged to doubt them she insisted on their truth 'with immovable conviction'. Needless to say, such an experiment would be unthinkable today.

The ease with which memories can be implanted gave rise to social mayhem in the 1980s and 1990s when many therapists adopted the view that psychological problems in adulthood could be traced to sexual abuse during childhood. Because of their traumatic nature, it was argued, such memories were repressed, and the main purpose of therapy was to recover these memories, so that patients could then face the real causes of their problems and deal with them—presumably with the therapist's help. The most extreme expression of this view was a book by Ellen Bass and Laura Davis entitled *The Courage to Heal*, which was first published in

1988 but has since gone through several editions. Bass and Davis, who had no formal training in psychology or psychiatry, were nevertheless bold enough to tell their readers:

> If you don't remember your abuse, you are not alone. Many women don't have memories, and some never get memories. This doesn't mean they weren't abused.

Elsewhere in the book, Bass and Davis write: 'If you think you were abused, and your life shows the symptoms, then you were.' This statement commits the logical fallacy of affirming the consequent. It is, of course, true that childhood abuse can cause later symptoms of psychological distress, but this does not mean that psychological stress must have resulted from childhood abuse. Murder results in death, but this doesn't mean that death is always due to murder. Unfortunately, widespread acceptance of Bass and Davis's edict too often led to aggressive therapy designed to help distressed people recover the memories of the abuse that led to their distress, when no such abuse had in fact occurred.

The problem was that therapists could easily and unwittingly implant false memories. Of course some people who suffer from psychological problems have indeed been abused sexually, or in other ways. But the notion that all or even most psychological problems result from abuse is almost certainly wrong, and innocent people were accused of abuses that they had not committed. This sorry era led to a good deal of research into the nature and fragility of memory, and a well-trained therapist should now be alert to the danger of implanting false memories, and of assuming abuse where other causes of mental distress are possible, and often more likely.

Memory is a fickle witness anyway, and decisions based on memory, whether in the courtroom or the therapist's office, are bound to lead to error. Sometimes the innocent are found guilty, and sometimes the guilty are declared innocent. The question then becomes one of determining the cost of a false decision. Which is the more costly: failing to detect a true criminal or true child abuser, or convicting innocent people of crimes or abuses they didn't commit? Based initially on Roman law, modern constitutions assert the right of those convicted to be deemed innocent until proven guilty. In the eyes of the law, at least, it is better to let at least some criminals walk free than to imprison innocent citizens. But often it is that arch-deceiver, memory, that is the true villain.

So why is memory so bad? It was clearly not designed by nature to be a faithful record of the past. Rather, it supplies us with information—some true, some false, and always incomplete—that we use to construct stories. 'Memory is a poet, not an historian', the American poet Marie Howe once said. We may well be what we remember, at least in part, but our memories, like clothes, can be selected and modified to create what we want to be, rather than what we actually are. Hillary Clinton, as the United States First Lady in 1996, told of visiting Bosnia, heroically risking her life as she disembarked from the plane and ran for cover under sniper fire. In fact, her landing was peaceful, and she was met by a smiling child, whom she kissed. Of course, she may have made up the story to seem heroic, but even if so, some commentators suggest that she actually came to believe her own account.

Ronald Reagan, too, recalled acts of heroism during World War II that seemed to derive from old movies. He even gave the impression he had been at the Normandy landings and at the liberation of Nazi death camps. Later, though, he came to accept that some of his reported adventures were not real, telling one of his associates: 'Maybe I had seen too many war movies, the heroics of which I sometimes confused with real life.'

Both Clinton and Reagan may have been lying, but it is more charitable to suggest that they were victims of self-deception. According to William von Hippel and Robert Trivers, the capacity for self-deception evolved because it reduces the chance of being found out. Lies told deliberately can often be detected, especially if the liar is well known to his or her audience—lie detectors don't work very well because they are not tuned to idiosyncrasies of the liar. It's easier to tell if your friend is lying, because of giveaway hesitations or unusual mannerisms, than to avoid being hoodwinked by a plausible stranger. But if the teller of false information believes that information to be true, it is then told with the equanimity of revealing the truth, and both the teller and the audience are deceived. People may come to really believe that falsely remembered events actually happened, the more they create vivid images of those events in their minds.

In any case, life might indeed be drab if all memories were accurate, and accurately told. The late Ulric Neisser, one of the giants of cognitive psychology, wrote: 'Remembering is not like playing back a tape or looking at a picture; it is more like telling a story.' And the story that it tells is as often directed to the future as to the past, as I'll explain in the next chapter.

3.
ON TIME

Time present and time past
Are both perhaps present in time future...
    —T. S. Eliot, from *Four Quartets*

Remembering is mind-wandering into the past. We can also wander into the future, imagining what might happen tomorrow, or next Christmas, or when the Antarctic ice melts. The evidence shows, in fact, that people spend more time thinking about the future than about the past. Nevertheless, there is a natural continuity between future and past, as time glides relentlessly from one to the other. What we're about to do quickly becomes what we have done—assuming we actually do it. Sometimes we don't, and when that happens we're inclined to say: 'Well, I forgot.' Even forgetting, it seems, can apply to the future as to the past.

Our ability to travel mentally into past and future, and the smooth continuity between them, underlies our sense of time itself. Although we can mentally travel in either direction, our physical lives are rooted to the present while time flows by. Downstream lies that singular event we all mercifully forget—or were incapable of remembering—called birth, while the lines of Isaac Watts' hymn 'Our God, Our Help in Ages Past' remind us of what lies upstream:

Time, like an ever rolling stream,
Bears all its sons away;
They fly, forgotten, as a dream
Dies at the opening day.

But although our actual lives are imprisoned between birth and death, we can mentally travel beyond both. History can be brought alive through past records and texts, or the discovery of ancient artefacts, and embellished in historical novels or movies. Futuristic scenarios can tell of brave new worlds, or impending disaster. Ray Bradbury's dystopian novel *Fahrenheit 451* depicts a future America where books are forbidden, and houses containing books are ordered to be burned. This dire outlook doesn't seem to have affected sales, although Bradbury himself is reported to have said: 'I wasn't trying to predict the future. I was trying to prevent it.'

Once we humans discovered the concept of time, we could then ask how far it could be stretched. Physicists tell us that a big bang 13.77 billion years ago started it all, and 7.5 billion years in the future the sun will grow so large it will gobble up the earth. These cataclysmic events, I think, take us well beyond the imaginable—well outside the limits of mental time travel, although I suppose we might well entertain the possibility of moving to somewhere in space where there is a less voracious sun. Dream on.

We rely very largely on our remembered pasts to construct our futures. Memory, in the form of knowledge as well as of remembered episodes, provides the building blocks from which to construct future plans. In the previous chapter, I referred to experiments in which people are asked to remember 100 episodes and identify a person, an object and a location. These experiments then continue as follows. We rearrange the remembered elements into new combinations, and our subjects are then asked to imagine future episodes built around them. For instance, a subject might remember her friend Mary dropping her laptop in the library, her brother Tom falling off his bicycle in the park, or her partner Shane cooking sausages in the kitchen. She might later be asked to imagine a future episode with her friend Mary cooking sausages in the park—an event that never happened, but is easily imagined. Our studies show that the areas in the brain activated by remembering these past events overlap extensively with the areas activated by the imagined future events. The brain hardly knows the difference.

People with amnesia typically have as much difficulty in imagining future events as they do in remembering past ones. Neither Henry Molaison nor Clive Wearing, whom we encountered in the previous chapter, could envisage future episodes any more than they could remember past ones. Deborah Wearing entitled her book on Clive *Forever Today*, and Suzanne Corkin called her book on Henry Molaison *Permanent Present Tense*; both titles capture the fact that both Clive and Henry had no sense of either past or future. Their minds were stuck in the present, with nowhere to wander. When Henry was once asked: 'What do you think you'll do tomorrow?' he replied: 'Whatever is beneficial.' Perhaps his inability to mentally wander into past or future relieved him of the worry that often plagues our mind-wanderings, and made him an exceptionally agreeable and cooperative subject.

Here's another patient with deep amnesia, known as 'N.N.', in conversation with the psychologist Endel Tulving:

E.T.: Let's try the question again about the future. What will you be doing tomorrow? (*There is a 15-second pause.*)

N.N.: I don't know.

E.T.: Do you remember the question?

N.N.: About what I'll be doing tomorrow?

E.T.: Yes. How would you describe your state of mind when you try to think about it? (*A 5-second pause.*)
N.N.: Blank. I guess.

When asked to compare his state of mind when he is trying to think about what he will be doing tomorrow with his state of mind when he thinks about what he did yesterday, N.N. described it as 'a big blankness' that was 'like swimming in the middle of a lake. There's nothing there to hold you up or do anything with.'

Many of the scenarios we envisage in the future, such as a dinner party, are based on past episodes, with some rearrangement to accommodate a new location, or a new combination of people. Perhaps this helps explain why memory for episodes itself is not always accurate. If we are to design futures based on our memories, we need our memories to be useful rather than accurate. By constructing possible futures, we can then select what seems the best plan—the most fun, perhaps, or the least likely to prove disastrous. In our mind's eye, we can imagine different scenarios for a wedding, say—where to hold it, who to invite, what music to play, even whether to go through with it. We play out different versions of a job interview, a new date, a tennis match, with the hope of figuring out the best strategy. The very flexibility of our memories can make for well-adjusted futures, but play havoc with the remembered past.

As children grow, their capacities to remember the past and imagine the future seem to surface together, somewhere between the ages of three and four. Neither capacity, though, comes about as a sudden dawning. Three-year-olds often seem unable to tell you what happened at nursery school or playcentre, or what might happen tomorrow, yet they learn things, such as new songs or games—even new words, some of which they shouldn't use. They may have a sense of things that happened, or that will happen, but lack the mental machinery to put together a coherent scenario. Work by Thomas Suddendorf and colleagues suggests, though, that by age four most children have the basic mental components to be able to construct a possible future event. It may be that language in younger children is not well enough developed, so they can't find the words to describe what they did or what they plan to do. This argument, though, can be reversed. Language itself is designed to convey the non-present, and perhaps doesn't really develop until the sense of time itself emerges. In evolution, too, some capacity for mental time travel may well have evolved before we gained the ability to talk about our mental travels, as I shall explore in the next chapter.

In the previous chapter, I suggested that we adapt our memories to create images of ourselves—politicians, for instance, seem especially prone to recalling acts of heroism that did not actually occur. We also create future images. William James, brother of the novelist Henry and regarded by some as the founder of scientific psychology, wrote of 'potential social Me' as distinct from 'immediate present Me' and 'Me of the past'. More recently, Hazel Markus and Paula Nurius write similarly of 'possible selves', based on how we see ourselves in the past but looking forward to new images of the self in the future. The idea of different possible selves provides much of the motivation that guides us as we plough through life. As Markus and Nurius put it: 'I am *now* a psychologist, but I *could* be a restaurant owner, a marathon runner, a journalist, or the parent of a handicapped child.' Future images can be both positive and negative—I can imagine myself as a roaring success, whether at parties, on the rugby field, or in scientific achievement, or I can see myself as a dismal failure at everything I do. Sigh.

Our imagined future selves can even extend beyond death. The ability to imagine beyond the reach of a lifetime reinforces religious belief, as we create for ourselves imagined heavens or hells. People can even be induced into bleak and self-destructive actions in their present lives through promise of a better life after death. Muslim children are taught from an early age that the main purpose of life in this world is to prepare oneself for an eternal and blissful life in the next, a promise that no doubt helped motivate the terrorists who flew planes into the Twin Towers in New York on 11 September 2001. The Japanese kamikaze pilots who died for their emperor in World War II may have similarly believed that they would be rewarded in the next life. Kamikaze means 'divine wind', and also refers to a cocktail made of triple parts vodka, triple sec and lime juice—a cocktail, perhaps, to die for. Various Christian sects have also indulged in flagellation, vows of poverty, or vows of silence, perhaps in the hope that they would lead more idyllic lives after death. The very notions of heaven and hell can be used to great effect to manipulate human behaviour, often for the benefit of kings and overlords. The offer of life after death, with associated rewards and punishments, is remarkably ingenious, since there seems no way in which we can be either gratified or disappointed—at least if these emotions are restricted to the living.

There may be less reason to believe in a life before one's own incarnation, since it has little consequence for the present life—although it may lead one to claim special qualities based on an earlier existence. One of my former classmates fervently believes he is the reincarnation of Benjamin Franklin. Reincarnation is a central tenet of Indian religions, as well as of a number of others, such as druidism and theosophy. Several Greek philosophers, including Plato, Pythagoras and Socrates, believed in reincarnation. In Buddhist philosophy, different incarnations can spread across six different realms of existence, including the human, the animal, and several kinds of superhuman existence. Only rarely, they say, is a person reborn in the form of a human. A possible exception, if my old classmate is to be believed, was Benjamin Franklin. These beliefs are all testimony to the inventiveness of mental time travel.

Many, if not most, of our activities are directed in one way or another to the future, but need not actually involve mental time travel. Instinctive behaviours evolved precisely because they increase the chances of survival, or the survival of our offspring—that's what evolution is all about. Even instinct, then, is future-oriented. We may flee from danger, fight the aggressor, eat the apple, flirt with the new neighbour, not because we imagine the consequences of doing so, but because we are driven by instincts of fear, anger, hunger, or sex. Much of our learning, too, is based on ritual, or what our parents think is good for us, rather than on our own imagining of our futures. But mental time travel goes beyond instinct and learning by providing flexibility, allowing us to play out options and check their likely consequences. We can mind-wander into the future to see what might happen.

This is not an argument against evolution. The capacity for mental time travel *itself* surely evolved through natural selection, but provides much more flexible and rapid adaptation to the contingencies of a complex world than is provided by the slow mechanisms of genetic change. Learning provides a faster means of adapting to what life throws at us, but it is still a slow train. We plod through school and piano lessons, learn habits and rituals, but even these are sluggish and inflexible compared with the ability to conjure up scenarios, and fine-tune our lives.

Is mental time travel unique to humans?

In his poem 'A Grammarian's Funeral', published in 1855, the English poet Robert Browning wrote: "'What's time? Leave Now for dogs and apes! Man has Forever!'" The idea that mental time travel, along with the concept of time itself, is peculiar to humans has been proposed by many, including Thomas Suddendorf and myself. Certainly, we humans seem obsessed with time. Events located in time weigh heavily on our conscious lives. We reminisce about the past, glorying in real or imagined triumphs, or regretting past mistakes. We dream about bright futures, vacations in the sun, or potential disasters. We are ruled by clocks, calendars, diaries, appointments, anniversaries—and taxes. We measure time on scales ranging from nanoseconds to aeons. Perhaps it's all a bit much, and we should heed the advice of Buddha and try to live more in the present, like Browning's dogs and apes.

Again echoing Browning, the German psychologist Wolfgang Köhler suggested that even our closest non-human relative the chimpanzee is, like Clive Wearing, stuck in the present. Köhler happened to be working at a primate research facility maintained by the Prussian Academy of Sciences in the Canary Islands when the First World War broke out. Marooned there, he occupied his time studying the behaviour of nine chimpanzees contained in a large outdoor pen. His work is famous for showing that chimpanzees are intelligent, and sometimes solve mechanical problems through the use of insight rather than mere trial and error. Köhler nevertheless concluded that, for all their problem-solving skills, chimpanzees had little conception of past or future.

Nevertheless, the idea that only humans are able to travel mentally in time, and imagine past and future events, is open to challenge, not least by a male chimpanzee called Santino in Furuvik Zoo in Sweden. Santino likes to collect stones and throw them at visitors. He gathers them well in advance of visitors arriving, and hides them so the visitors won't see them. It is difficult to avoid the impression that Santino is planning a specific future event, perhaps even seeing it in his mind's eye as he gleefully stockpiles his ammunition. Santino is not alone. In his book *The Descent of Man*, Charles Darwin tells of a baboon at the Cape of Good Hope who threw missiles at people and prepared mud in advance for the purpose. The stockpiling of missiles is also a characteristic of that other dangerous primate, *Homo sapiens*. According to the Federation of American Scientists, Russia has 4650 active nuclear warheads, while the US has a mere (but more mathematically pleasing) 2468.

It's not just apes; even birds seem to show evidence of mental time travel. Clark's nutcrackers conceal items of food in thousands of locations, and later retrieve them with remarkable, but not perfect, accuracy. Scrub jays also cache food, and experiments suggest that they remember not only where they have cached it, but also which items are stored in which locations, and when they stored them. For example, if they have cached worms and peanuts, they will retrieve the worms if given the opportunity quite soon after caching them, because worms are more palatable than peanuts, at least to the jays. But if there is a delay, they will prefer the peanuts, because the worms will have decayed and become inedible. This has been taken to mean that they remember the act of caching in enough detail to know what they cached, where they cached it, and when they cached it. A simpler explanation, though, is that they mentally tag each item with a 'use-by date', and so know how long ago it was cached, rather than remembering the actual act of caching.

They also seem to cache with a future event in mind. If they are watched by another jay while caching, they will often re-cache the food in different locations when the watcher is no longer present. They evidently fear the possibility that the watcher will later steal the food; although it takes a thief to know a thief—they will only re-cache if they have themselves stolen cached food. Further, when given a choice of food to cache, scrub jays choose not in terms of their present hunger but in terms of what food they will want to eat the following day—anticipating breakfast, in other words.

Similarly, orang-utans and bonobos have been shown to save tools not needed right now for use up to 14 hours later. Some groups of chimpanzees store hammers and anvils for years of use in cracking nuts. Tool-making itself can be taken as evidence for mental time travel into the future. New Caledonian crows make tools from twigs and bits of wire to solve mechanical problems. In some cases this may be a simple matter of improvisation to solve an immediate problem, rather than planning for a more distant future. Other examples, though, do suggest specific planning for future use. The crows carefully shape the leaves of pandanus trees for the specific purpose of extracting grubs from holes. Using their beaks, they taper the leaves to be wider at the end which is held in the beak and narrower at the end inserted into the hole. The birds choose pandanus leaves because they have angled spikes along one side, and these attach to the grubs so the bird can then pull them out. The making of these tools suggests meticulous planning. Not to be outdone by crows, chimps fashion sticks for fishing termites out of holes, and spears for plunging into the hollow trunks of trees to extract bushbabies, which they then eat. One colony of chimps uses tool sets of up to five different stick and bark implements to extract honey from hives.

In all such examples, though, we cannot be sure that animals are truly travelling mentally in time, envisaging past or future events. It is often possible to account for what looks like mental time travel in a bird or a chimp in terms of instinct or habit. In birds, for example, the caching of food is instinctive, although it may be modified by experience, as in the case of the scrub jays re-caching their food after being watched by a potential thief. And even re-caching may be the outcome of a learned association between the presence of the thief and the subsequent loss of the cache, and need not imply actually envisaging a future theft. Tool-making industries among chimps may be the result of trial and error rather than planning, and passed between generations without any specific imagining of a future event. We humans learn lots of complex things, like reading or playing the piano, often without any conscious sense of what future it will bring.

Future-directed behaviours can be purely instinctive. Every year, Canada geese migrate south, in their distinctive V-formation, to escape the bitter northern winter. Some even reach Auckland, but there is no suggestion that they envisage the delights of New Zealand's 'most liveable city' before they take off. In this they differ from the Canadians themselves who migrate to Florida or Hawaii, no doubt in pleasurable anticipation of what they will find when they get there. Instinct alone can drive remarkably complex behaviours, from the making of dams, nests, or even spiders' webs to elaborate courtship rituals. Such activities are geared to future survival, but do not depend on mental travels through time.

Psychologists and ethologists have often reminded us to be careful when attributing human-like thoughts to animals. The English ethologist Conwy Lloyd Morgan, who had studied under Darwin's colleague and advocate Thomas Henry Huxley, famously established what has become known as Morgan's canon:

> In no case may we interpret an action as the outcome of a higher mental faculty, if it can be interpreted as the exercise of one which stands lower in the psychological scale.

The canon was declared in 1894, but ten years later the famous case of Clever Hans came as a salutary reminder. Clever Hans was a horse, and seemed able to answer complex questions by tapping a front hoof. When asked 'What is $\frac{2}{5}$ plus $\frac{1}{2}$?' he stamped his hoof nine times, paused, and stamped another ten times, apparently indicating that the answer was $\frac{9}{10}$. When asked a person's name, he would laboriously tap it out letter by letter, with one tap for A, two taps for B, and so on. Professor Carl Stumpf of the University of Berlin, a leading psychologist, was convinced of the horse's genius, until one of his students, Oskar Pfungst, showed that

Clever Hans was actually responding to subtle signals, given by his trainer, as to when to stop tapping. The trainer himself apparently did not realise that he, and not Clever Hans, was generating the answers.

The canon, also known as the principle of parsimony, has understandably been aimed at the idea that animals might be capable of travelling mentally in time. But there is also an uncomfortable sense that parsimonious explanations may serve to underestimate the intelligence of our animal cousins, and help preserve our human assumption of lordly superiority. The Bible gives added encouragement, as in the Eighth Psalm:

> What is man, that thou art mindful of him . . .?
> For thou hast made him a little lower than the angels,
> and has crowned him with glory and honour.
> Thou hast made him to have dominion over the works of thy hand;
> Thou hast put all things under his feet:
> All sheep and oxen, yea, and the beasts of the field;
> The fowls of the air, and the fish of the sea . . .

The main problem in deciding whether non-human animals have human-like thoughts is that only humans have articulate language. Language itself is a 'higher mental faculty', to use Lloyd Morgan's expression, and indeed some philosophers and linguists have maintained—wrongly, I think—that the very act of thinking depends on language. But whether or not this is so, we can discover a lot about thought by simply asking people to tell us of their experiences and thoughts. People have no difficulty describing their mental time travels—their memories, plans and fantasies. But even our closest non-human relatives, chimpanzees and bonobos, cannot actually tell us what's on their minds. Neither can the rather talkative parrot.

Maybe, though, there is another road into the animal mind that can help, and suggest that animals may indeed be capable of mental time travels. To explain this, I need to introduce you to another animal (or two).

4.
THE HIPPO
IN THE
BRAIN

You may think that there are other more important differences between you and an ape, such as being able to speak, and make machines, and know right from wrong, and say your prayers, and other little matters of that kind; but that is a child's fancy, my dear. Nothing is to be depended on but the great hippopotamus test.

—Charles Kingsley, from *The Water-Babies*

Kingsley was not referring to that large animal, but to a small structure in the brain known as the hippocampus minor. This structure was of intense interest following the 1859 publication of Darwin's *Origin of Species*, because the distinguished anatomist Richard Owen maintained that only humans possessed it. This showed, he said, that humans could not be descended from apes, as Darwin's theory implied. Darwin, normally mild-mannered, didn't much care for Owen, once describing him as 'spiteful, unfair, ungenerous, extremely malignant, false, rude, unjust, illiberal and disingenuous'. In defence of Darwin's theory, Thomas Henry Huxley, also known as 'Darwin's bulldog', showed Owen to be wrong by demonstrating that all apes in fact do possess a hippocampus minor. Although a minister of the church, the reverend Charles Kingsley was one of the first to praise Darwin's book, and his satirical reference to the hippopotamus was a gibe at the esteemed Dr Owen.

But Kingsley himself seems to have been a little confused. Following the above passage, he went on:

If you have a hippopotamus major in your brain, you are no ape, though you had four hands, no feet, and were more apish than the apes of all aperies. But if a hippopotamus major is ever discovered in one single ape's brain, nothing will save your great-great-great-great-great-great-great-great-great-great-great-greater-greatest-grandmother from having been an ape too.

Well, the confusion between hippocampus and hippopotamus was no doubt deliberate and intended satirically, but it was the hippocampus minor, not the hippocampus major, that was supposedly the critical structure. But perhaps he was prophetic, as we shall see.

The hippocampus minor soon disappeared as a serious contender for human uniqueness, or indeed for anything else, and even lost its name. It is now known by its original name of 'calcar avis', meaning cock's spur. In an entertaining article on the acrimonious debates between Owen and Huxley, Charles Gross remarks that after the controversy had subsided, the calcar avis was to be found 'only in obscure corners of human anatomy texts, where it still rests'.[*]

Of much more interest, as unintentionally foreseen by Charles Kingsley, is the hippocampus major, now better known simply as the hippocampus with the demotion of its one-time junior partner. It's not so much a hippopotamus that we house in our brains as a seahorse, since hippocampus means 'seahorse' in Greek. It was named for its resemblance to that undulating creature, in which the male carries the young (in case you didn't know). It is a structure

---

[*] In a search of the Web of Science, I could find nothing on what the calcar avis might actually do, although I did learn of a species of plankton known as *Pseudosolenia calcar-avis*, which is perhaps lurking to tip us humans from our pedestal.

on the inner surface of the temporal lobes of the brain—roughly behind your ears.

It is the hippocampus that is utterly critical to mental time travel—our ability to travel mentally backward and forward in time. The common feature of the amnesic cases Henry Molaison and K.C. and Clive Wearing, introduced if you remember in Chapter 2, is that all had suffered major loss of tissue in the hippocampus. In her book on Clive, Deborah Wearing records that she was shown a scan of his brain, and writes that 'by the time they had figured out what was wrong with Clive and started pumping the anti-viral drugs into him all he had left were seahorse-shaped-scars where his memory used to be'.

Figure 4.1. The hippocampus (*left*) and the real seahorse (*right*).

And it is the hippocampus that is at the hub of the system that lights up when people wander mentally back and forth in time. In preceding chapters, I also mentioned the Cluedo-like experiments in which we ask our subjects to recall 100 events in their lives, and then rearrange the people, objects and places featuring in these events and ask our long-suffering participants to imagine future scenarios based on these novel arrangements. The subjects lie in the MRI scanner while they perform these feats, and the areas of the brain that are activated correspond largely to the default-mode network. This is the 'mind-wandering' network, and includes prefrontal lobes, temporal lobes and parietal lobes. It matters little whether the participants are recalling the past or imagining future events—the activated areas overlap quite extensively.

The hippocampus is the Grand Central Station of this network, reciprocally connected to other areas in the network, including both the cortical areas above and the more emotional areas below. It is responsible for what has been called 'temporal consciousness', or knowing where you are in the ribbon of time. Oddly enough, although people with damage to the hippocampus seem lost in time and stuck in the present, they may still be able to answer questions about events in time that do not involve themselves—such as the death of Princess Diana, or what the next medical breakthroughs are likely to be. The job description of the hippocampus, it seems, is to deal with personal matters—the recording and retrieval of personal events, and the making of personal plans.

The hippocampus seems to be a forward-looking structure, with the front (anterior) end concerned more with the future and the rear (posterior) end with the past. In our Cluedo study, when people imagined future scenarios, and were later asked to try to remember them, both ends of the hippocampus were often activated. That is, imagined scenes were also remembered as though they had actually occurred. This could perhaps help explain why some memories are false—for instance, why Hillary Clinton remembered running to escape gunfire when she arrived in Bosnia, when in fact her arrival was peaceful and welcoming. Maybe she had imagined the threatening scenario before arriving there, and this memory had lodged in her brain as though it had actually happened. But who knows? Possibly Hillary herself doesn't know.

Besides its role in mental time travel, the hippocampus has another talent. It records locations in space. In 1978, John O'Keefe and Lynn Nadel (one-time PhD classmates of mine) wrote a book that has become a classic of neuroscience, called *The Hippocampus as a Cognitive Map*. Their work was based on the recording of activity from microelectrodes inserted into different regions of the hippocampus of the rat. If the rat was placed in a maze, the location of activity corresponded to where the animal was located. The single cells (or neurons) became known as 'place cells'—a bit like a GPS system lodged in the brain itself.

It turns out that our own human hippocampus also contains place cells. In a report published in 2003, neurosurgeons inserted electrodes in the hippocampi and other brain areas in patients being monitored for potential surgery for intractable epilepsy. Their purpose was to locate the seizure foci, but the electrodes also allowed them to record from single cells while the patients explored and navigated a virtual town on a computer screen. Some of the hippocampal cells responded to specific locations in the virtual town. Cells in the adjacent parahippocampal region also responded to views of landmarks in the town.

The hippocampus is not a static map, though. Activity in place cells adjusts when the animal, rat or human, moves into a new environment. Maps can also exist at different scales, much like the zoom function on internet maps. For instance, it appears that small-scale maps are located in toward the rear of the hippocampus, and large-scale maps toward the front. The coding of time is also graded, like an adjustable calendar. You can replay your past or imagine your future over years, days, or minutes. The representation of space-time in the hippocampus and neighbouring regions is complex, and not yet fully worked out.

The hippocampus also seems to swell to meet spatial demand. London taxi drivers have to undergo extensive training to learn the precise geography of that large and confusing city. They must be able to decide the quickest route to a passenger's destination immediately, without looking at a map, consulting a GPS system, or asking a controller by radio or cellphone. This requirement was set up in 1865, and is known as 'The Knowledge'. Brain-imaging has shown that the hippocampi of these taxi drivers are unusually enlarged. They are also larger than those of London bus drivers, who simply have to follow designated routes. London bus drivers, though, are better than the taxi drivers at learning new spatial tasks, suggesting that the taxi drivers may already have crammed about as much spatial information into their hippocampi as those little seahorses can take. In any event, it seems that the hippocampus is as important to knowing where you are in humans as it is in rats.

In rats, as in humans, the hippocampus also plays a critical role in laying down memories. It has been known for some time that if you stimulate a cell in the hippocampus with a high-frequency volley of electrical pulses, the connection (synapse) between that cell and the upstream one it connects to is strengthened. The effect is known as 'long-term potentiation', and is long-lasting, sometimes persisting for months. It was originally demonstrated in the rabbit in 1966 by Terje Lømo in Oslo, Norway, but has since been widely studied in rats, as in other species. It is generally considered to be the basis of memory. Your memories, then, are established through the strengthening of connections in your brain, with the hippocampus playing the commanding role. This is not to say that memories are lodged in the hippocampus alone. Long-term potentiation may hold them there for a while, but they eventually diffuse into other regions of the brain. And it's the hippocampus that finds them again.

We should not be surprised that the hippocampus is involved in recording where we are in space as well as in time, since mental time travel takes place in a space-time manifold. It is, as I suggested earlier, the Grand Central Station for our mental excursions, recording our mental comings and goings. But its seemingly similar roles in rat and human open the possibility of answering the question I raised in the previous chapter: is mental time travel unique to humans?

The secret life of Walter Ratty

In their classic 1978 book, O'Keefe and Nadel wrote that the addition of a temporal component 'changes the basic spatial map into a human episodic memory system'. The question has since arisen, though, as to whether the temporal component was already present in our mammalian forebears. Recent evidence suggests that even rats may imagine past and future events.

Place cells in the rat hippocampus are occasionally active after the rat has been in some specific environment, such as a maze, as though the animal is actively remembering where it was, or perhaps imagining where it will be—or might be. This activation occurs in what are known as 'sharp-wave ripples' which sweep out sequences of place cells, as though the animal is mentally tracing out a trajectory in the maze. This occurs sometimes when the animal is asleep, or when it is awake but immobile. It is as though the animal is replaying its experience in the maze, perhaps while dreaming or daydreaming—for a laboratory rat, being in a maze is probably the most exciting event of its day. The ripples, then, suggest that the rat is mentally wandering from one part of the maze to another.

These mental perambulations need not correspond to the paths that the rat actually traversed. Sometimes the ripples sweep out a path that is precisely the reverse of one the rat actually took. It may be a path corresponding to a section of the maze the rat didn't even visit, or a short cut between locations that wasn't actually traversed. One interpretation is that the ripples function to consolidate the memory for the maze, laying down a memory for it that goes beyond experience, establishing a more extensive cognitive map for future use. But mind-wandering and consolidation may be much the same thing. One reason that we daydream—or even dream at night—may be to strengthen our memories of the past, and allow us, and the rat, to envisage future events. I return to the dream world in Chapter 7.

The imagined trajectories, if that is what they are, are more rapid than the actual ones. I think this is true of our own mental wandering. It takes me about an hour to walk to my place of work from my home, but when I imagine the walk, and the landmarks I encounter, it takes less than a minute or so. Mentally, we travel in the fast lane. It is not altogether clear, though, whether time itself moves faster in the mental world, or whether we flit from location to location, leaving out chunks of the journey.

In another clever experiment, rats were placed in an environment with 36 food wells, arranged in a six-by-six array. They were given experience eating from these wells, so they learned the environment pretty well. One particular food well was then established as a Home location where food could be found, and the rats were then lured to different locations from which they had to find their way Home. The researchers recorded from multiple sites in the hippocampus, and discovered ripples corresponding to these Home routes, but they were played before the rat actually embarked on the journey. What is interesting is that they were generally routes the animals had not actually traversed before. This seems to be mental time travel into a future event. The authors of the study suggest that the hippocampus 'functions in multiple conceptual contexts: as a cognitive map in which routes to goals might be explored flexibly before behaviour, as an episodic memory system engaging in what has been termed "mental time travel" . . .'. The hippocampus, in other words, can lay out an action plan.

Perhaps to labour the point, recordings from the hippocampus seem also to tell which way a rat will turn at a choice point in a maze. The rats in question were trained to alternate left and right turns at a particular spot in the maze. Between trials, they were taken out of the maze and placed in a running wheel. While they were running, recordings from their rippling hippocampi again revealed activity corresponding to paths taken in the maze, including which way they would turn when next placed in the maze. The rats, it seems, were planning their next turn. My mind wanders, too, when I'm on a treadmill, but I also use time on the treadmill to figure out what I might do later. How do we know the rat was not just reminiscing about the way it turned on some previous trial, and not on what way they will turn next? Well, perhaps because of the time they spent in the running wheel, they sometimes made errors when placed back in the maze—say, turning left instead of right. But this error was signalled by the hippocampal activity, showing that the rat was actually planning a wrong turn. The study's authors write that activity in the hippocampus, 'having evolved for the computation of distances, can also support the episodic recall of events and the planning of action sequences and goals'.

Maybe hippocampal recordings might one day help goalies know in advance which way the kicker will shoot when given a free kick at the goal.

I have dwelt on these rat experiments because they seem to show that even the humble rat indulges in mental time travel. Like rats, we are creatures that move on the face of the earth, so space is fundamental to our wanderings, both physical and mental. It would not be surprising, then, if our mental time travels did evolve from the simple replaying, and pre-playing, of spatial movements. You have to go back some 66 million years to find the common ancestor of rat and human. Over that long interval our mental faculties have surely diverged, but in a spatial world the mechanisms involved in living, remembering and planning in space are critical, and are probably conserved through evolutionary time. Mental time travel may well be one of the earliest of mental faculties to evolve.

It is fundamental to all moving animals to know where they are, where they've been, and where they're going next. In Chapter 2, I described the mnemonic device known as the method of loci, whereby we remember lists of things by locating them mentally in some familiar terrain, and then mentally wander through the terrain to recover them. This no doubt derives from our spatial heritage.

What about birds, who are surely the experts in spatial travel? They indulged in air travel long before we humans did, and still do it much more gracefully than we do in our clumsy air machines. For a while it was thought that birds did not have hippocampi, leading to the frivolous theory that the function of the hippocampus was to prevent flight. But the bird brain is organised rather differently from the mammalian one, and it turns out that there is a region in the bird brain that is homologous to the mammalian hippocampus. It develops from a region of the embryo that corresponds to the part of the mammalian embryo from which the mammalian hippocampus derives. Anatomists now identify this region as the avian hippocampus. Far from preventing flight, the avian hippocampus is critical to their travel plans, as well as their food-gathering strategies. Not surprisingly, birds that cache food in multiple locations have larger hippocampi than those that don't. In this, they are the avian equivalents of London taxi drivers.

Of course, our own mental travels are more complex than simply moving from place to place. For a start, our cognitive maps are extraordinarily flexible. As I suggested earlier, they can zoom. Let me take you on a brief tour. First, imagine you are sitting at your desk (as I am right now). You can imagine the other objects on the desk—a half-finished crossword, a small stack of books, an empty cup. Zoom back a little and imagine the room, the sofa, the bookcase lining the far wall, the door leading to the hallway. Zoom back further and take a mental wander around the house (or apartment). Zoom out now to the suburb, the small row of shops, the bus stop, the intersection of streets. Take a deep breath and keep zooming—to the city, the country, the world. You can also flit about—to Paris, New York, or the place with the forgotten name in the Italian Alps.

You can also link these many locations with times, albeit imprecisely. Location, after all, is time, since you can only be in any one location at any one time. You can zoom in time as well, looking back or forward over seconds, minutes, hours, days, weeks, months, years, decades. And all of these mental journeys through time and space are populated with people, events, things, emotions, disappointments, triumphs—the rich fabric of our lives. And although we forget much of what has happened in our lives, we do remember a lot—enough to write biographies or bore our colleagues and children. Our plans are likewise richer than simply taking a new route to work. And then there is the world of fiction, the made-up stories and fantasies that occupy much of our conscious lives—but more of that in Chapter 6.

When Thomas Suddendorf and I framed the idea of mental time travel, we also outlined additional mental resources needed to construct an imagined episode, whether past or future. We might need an executive processor to build the episode from its constituent parts, a memory buffer to hold information before it fades away. In the previous chapter, I referred to work by Suddendorf and associates implying that children cannot construct fully coherent episodes of past events until they're about four years old. It may well be stretching credulity to suppose that Walter Ratty has the mental machinery of even a four-year-old child.

But do these qualities really distinguish our mental travels from those of other creatures? We need to be wary, on the one hand, of the 'Clever Hans' error of attributing implausibly human-like qualities to non-human species, but on the other hand of building impregnable mental fortresses that no animal could invade. In a prescient discussion in 2006 on the possible role of the hippocampus in episodic memory, David Smith and Sheri Mizumori wrote:

> We leave it to others to debate whether rodents possess the capacity for consciousness and mental time travel. In any case, the history of psychology is replete with examples of 'uniquely human' cognitive functions, which were later demonstrated in so-called lower animals. Given the remarkable homology of mammalian nervous systems and the fact that the ability to explicitly recall previous experiences has such obvious adaptive value, we suggest that, in the absence of contradictory evidence, the most conservative position is to assume that rodents possess an episodic memory system that is qualitatively similar to that of humans.

Charles Kingsley would surely have applauded. And that other Charles, in *Origin of Species*, famously wrote: 'The difference in mind between man and the higher animals, great as it is, certainly is one of degree and not of kind.'

But there is one place that may be beyond the limits of Walter Ratty's mental travels, and that may still tell us something about our own mental wanderings that Walter never dreamt of. That comes next.

5.
WANDER
ING INTO
OTHER
MINDS

• • •

Imagining what it is like to be someone other than yourself is at the core of our humanity.

—Ian McEwan, from 'Only Love and Then Oblivion'*

In the quote that opened this book, as I noted, it was not really Walter Mitty whose mind was wandering. The wanderer was, of course, James Thurber himself, who had ventured into the mind of a fictional character, and set that character mind-wandering on a dangerous mission. In everyday life, as in fiction, we often assume the identities of others. A good actor can transport herself into another person, and carry an audience along with her. Even TV soap operas can take us into other families or situations, and allow us to identify with imaginary people. We habitually judge the personalities of others, try to figure out how they think and act, perhaps to decide whether to employ them, consult them, or marry them.

The sense that we know what others are thinking has often led people to believe in psychic powers, or telepathy, as though minds can communicate without any contact through the senses. In Chapter 1, I mentioned the German physician Hans Berger, whose fall from a horse was sensed by his sister who was kilometres away and could not have seen or heard the event. Berger thought this might have been an instance of telepathy, but his own attempt to prove an electrical basis for telepathy failed. Nevertheless, many distinguished people have firmly believed that thoughts can be transferred by non-physical means, and even

* An essay published in *The Guardian*, 15 September 2001.

that we can communicate telepathically with the dead—or they with us.

The idea seems to have been especially popular in late nineteenth-century England. In 1882, the Society for Psychical Research was established in London to investigate telepathy and other so-called psychic phenomena, such as ghosts, trance states, levitations, mediums and communication with the dead. Its first president was Henry Sidgwick, later Professor of Moral Philosophy at Trinity College, Cambridge, and other distinguished members included the experimental physicist Lord Rayleigh, the philosopher Arthur Balfour, who became Prime Minister of England from 1902 to 1905, and Sir Arthur Conan Doyle, author of the Sherlock Holmes stories. The Society attracted famous psychologists such as Sigmund Freud and Carl Jung, and the American psychologist William James was so impressed that he established the American Society for Psychical Research.

Many people still believe in telepathy, also known as extrasensory perception, or ESP. A 1979 survey of over 1000 academics in the United States showed that 55 per cent of natural scientists, 66 per cent of social scientists and 77 per cent of those in the arts, humanities or education believed that ESP was either established or a likely possibility. Of course, academics will believe almost anything. Psychologists, though, are the spoilsports; for them, the equivalent figure was only 34 per cent, and an equal number believed ESP to be impossible. I doubt that these figures have changed much, or that the beliefs outside of the US are likely to be much different. The stumbling block is that ESP implies an effect that operates at a distance, but without any clear physical medium, such as light, sound, smell, or even radio waves, and this seems both physiologically and physically implausible, or downright impossible.

Well, maybe not. Some have appealed to the fundamental nature of physical reality. A theorem attributed to the British physicist John Stewart Bell says that any model of reality consistent with quantum mechanics must be non-local. That is, any particles that have once interacted can become entangled, such that when they are later separated, observations on one of the particles can affect what will be observed on its entangled partners. This is true no matter how far apart they are, and is incompatible with any physical signal from one to the other. You might think that this also applies to people, perhaps especially those who have had multiple entanglements with people they know or knew well, such as lovers or marriage partners, and so explains ESP.

In his 2006 book *Entangled Minds: Extrasensory Experiences in a Quantum Reality,* the parapsychologist Dean Radin, a one-time engineer and concert violinist, has indeed argued that the physical interactions at a distance implied by Bell's theorem could explain ESP (also known as psi). He concludes:

> ... over the past century, most of the fundamental assumptions about the fabric of physical reality have been revised in the direction predicted by genuine psi. This is why I propose that psi is the human experience of the entangled universe. Quantum entanglement as presently understood in elementary atomic systems is, by itself, insufficient to explain psi. But the ontological parallels implied by entanglement and psi are so compelling that I believe they'd be foolish to ignore.

Maybe if Hans Berger had known more about particle physics, he would have performed a different experiment, but then we would have been denied the discovery of electroencephalography, and perhaps even of the wandering mind.

As a psychologist, and one who keeps a watchful eye on the field and occasionally tries to make intellectual contact with entangled particles, I remain firmly sceptical. It seems unlikely that entangled particles have anything to do with the human mind. Thousands of experiments have been conducted to test the existence of the paranormal, but the published evidence is unconvincing, especially if one considers that most negative results remain unpublished. Indeed, I was once confronted by a group of students who resented my scepticism, and between us we set up a study, but the results were relentlessly negative—and of course remained unpublished (until now).

Given the wish to believe in disembodied minds, fraudsters have been quick to cash in. One famous example is Uri Geller, an Israeli-British stage performer, who rose to fame in the 1970s for television shows on which he claimed to demonstrate psychic powers. He is perhaps best known for his prowess at bending spoons, apparently through the power of thought—a phenomenon which, if true, is an example of psychokinesis. Geller's feats are easily duplicated by sleight of hand and spoon, without resort to psychic powers, by stage magicians, including James Randi, who wrote a book entitled *The Magic of Uri Geller*—it was later called *The Truth About Uri Geller*. The James Randi Educational Foundation was established in 1996 to further Randi's work. It offers a $1,000,000 prize to anyone who can demonstrate psychic powers (see www.randi.org if you want to challenge)—and to this day it has not been collected.

Geller's exploits were also unmasked, leaving no spoon unbent, in New Zealand by two psychologists, David Marks and Richard Kammann, who were able to repeat his demonstrations on television, again without any claim to psychic powers. They too wrote a book exposing the field of psychic phenomena, and Geller in particular, entitled *The Psychology of the Psychic*. I recommend these books, but alas they do not have the selling power of books that proclaim the existence of the psychic.

In spite of the lack of evidence, we humans do seem naturally inclined to believe in powers of the mind to transcend physics, whether through ESP, clairvoyance, telekinesis, or communication with the dead. In part this is probably a matter of wishful thinking. It is comforting to think that the minds of the dead live on, and that we can still communicate with them—or indeed that our own mental lives will soar on after death, unencumbered by the now useless body. The psychologist Paul Bloom, in his book *Descartes' Baby*, goes so far as to suggest that we are actually born to be philosophical dualists, like Descartes himself, believing the mind to be separate from the body. Dualism, Bloom suggests, is hard-wired.

This is not to say, of course, that our minds are in fact separate from our bodies—it's just that we're predisposed to believe so. It is indeed difficult to convince most people, except us doughty psychologists and materialistic neuroscientists, that we are merely creatures of flesh and bone, with physical processes inside our heads that dictate our thoughts and actions. The belief in dualism, the idea that the mind can escape the body and the constraints of the physical world, is one facet of mind-wandering itself.

Theory of mind

Whether or not the mind is actually constrained by the mechanical functioning of the brain, we are actually very good at knowing what others are thinking, a talent known as 'theory of mind'. There is no compelling reason to believe that this is due to non-material influences, such as entangled particles. It is partly intuitive, based on subtle cues we may not be aware of, but that are nonetheless received via the senses. It is partly due to cultural sharing. People of the same culture tend to respond in the same ways to the same situations—we tend to be embarrassed by the same social blunders, elated by the same victories, saddened by the same losses. And we share senses—we see what others see, hear what they hear, smell what they smell. We even share our mind-wanderings through storytelling, although that itself is a story for the next chapter. We also use simple observation to infer what's going on in the minds of others.

This is nicely illustrated by the Sally–Anne test, which is a test of children's ability to infer that other people have a false belief about something. The child is shown a scene involving two dolls, Sally and Anne. Sally has a basket and Anne has a box. Sally puts a marble in her basket and leaves the scene. While Sally is away, naughty Anne takes the marble out of the basket and puts it in her box. Sally then comes back, and the child watching all this is asked where she will look for her marble. Children under the age of four typically say she will look in the box, which is where the marble actually is. Older children will understand that Sally did not see the marble being shifted, and will correctly say that Sally will look in the basket. They understand that Sally has a false belief. To that extent, at least, they know what's in Sally's mind, and that this is different from what's in their own minds.

Curiously enough, children younger than four seem to act as though they understand what others believe even if they can't say so. In a remarkable Hungarian study, babies as young as seven months were influenced by the belief of another individual. The babies were shown movies of a ball that rolled behind a screen. The ball could then either stay behind the screen or roll away. The events were also watched by a cartoon character, who sometimes left the scene and came back. The position of the ball could be changed while the character was away, so he could believe the ball to be behind the screen when it wasn't, or not behind the screen when it was. When the character returned and the screen was removed, the babies looked longer at the scene when the cartoon character's expectation was not met. That is, they seemed to expect the cartoon character to be surprised, as though they could read his mind.

This experiment suggests that an understanding of others' beliefs can influence the way even young babies act, even if they can't put words to that understanding. In the same study, adults were found to behave in very much the same way, with their actions influenced both by their own beliefs and those of another observer. The authors of this study write:

> The finding that others' beliefs can be similarly accessible as our own beliefs might seem problematic for an individual, because it may make one's behavior susceptible to others' beliefs that do not reliably reflect the current state of affairs. However, the rapid availability of others' beliefs might allow for efficient interactions in complex social

groups. These powerful mechanisms for computing others' beliefs might, therefore, be part of a core human-specific 'social sense', and one of the cognitive preconditions for the evolution of the uniquely elaborate social structure in humans.

This social sense seems to be acquired at a very early age, and may even be inborn.

To find out which parts of the brain are activated in mind-reading, people are placed in a brain scanner and told stories that allow them to deduce the beliefs of others. One story, for example, describes John telling Emily that he drives a Porsche, but his car is in fact a Ford. Emily knows nothing about cars, and so she believes that John's car is a Porsche. Emily then sees the car, and the person in the scanner is asked what Emily thinks is the make of the car. Most people understand that Emily falsely believes it to be a Porsche. Understanding that others have false beliefs again activates the default-mode network, suggesting that mind-wandering can indeed take us into the minds of others.

The understanding that others can have beliefs different from one's own seems to be the most telling example of theory of mind. It is critical to social harmony, and allows us to correct mistaken beliefs in others—or at least try to do so. Emily, for example, might usefully be told that John is something of a liar, and shouldn't be trusted. It can work both ways, though, since John might usefully, if cruelly, be told that Emily knows nothing about cars so he can create in her a false belief that is to his advantage—although probably only a temporary one. In a tolerant society, what is important may not be the understanding that other beliefs are false, so much as the understanding that people can have widely differing beliefs. Many, I am sure, will reject my belief that ESP does not exist, but understand that I hold that belief, and possibly feel sorry for me.

Theory of mind has a recursive property, such that understandings can be embedded in understandings. I may believe that you believe that I believe in Santa Claus. Or I may believe that you feel sorry for me because you believe that I don't believe in ESP. The psychologist David Premack takes it even deeper, suggesting: 'Women think that men think that they think that men think that women's orgasm is different.' Well, he's a mere man, and it is he who thinks *that*, taking it deeper still. Runaway recursion may have been driven by the human propensity to deceive. If I know what you believe about me, I can then deceive you by acting contrary to that belief. This is well illustrated by the old Yiddish joke about a man who meets a business rival at a train station and asks where he is going. The business rival replies he is going to Minsk. The first man then says, 'You're telling me you're going to Minsk because you want me to think you're going to Pinsk. But I happen to know that you are going to Minsk, so why are you lying to me?'

As the English poet and novelist Sir Walter Scott[*] put it: 'O what a tangled web we weave, / When first we practise to deceive!'

People do vary, though, in the capacity to read minds. At one extreme is what has been termed 'mind-blindness', said to underlie the condition known as autism. One well-known case is a woman called Temple Grandin, who has a PhD in agricultural science and works at Colorado State University. Her autism doesn't seem to

---

[*]   No, it wasn't Shakespeare. The quote is from Scott's epic poem *Marmion*.

have affected her intelligence, as she has written several books, three of them describing her own condition. Lacking a natural social understanding, she was forced to resort to detailed observations of people's actual behaviour in order to figure out how to behave appropriately in social situations. The habit of close observation paid dividends, though, in her work on animal behaviour. The title of her most recent book is *Animals in Translation: Using the Mysteries of Autism to Decode Animal Behavior*, which prompted a BBC documentary unkindly entitled *The Woman Who Thinks Like a Cow*.

High-functioning autism, as evident in people like Temple Grandin, is also known as Asperger syndrome. People with this condition often can pass false-belief tests, such as the Sally–Anne test, but they apparently do so only through verbal reasoning and explicit instructions about the task. As I noted earlier, normal infants seem instinctively to demonstrate an understanding of false belief well before they can demonstrate it verbally, since they will look to where an actor mistakenly believes an object to be hidden. People with Asperger syndrome do not do this, suggesting that the spontaneous understanding of false belief is lacking.

People like Temple Grandin lie at one end of what is actually a continuum. Some have suggested that people at the opposite end of that continuum have an obsessive sensitivity to what others are thinking, perhaps leaning to paranoia and magical thinking, and even schizophrenia.

In his aptly titled book *Knots*, the Scottish psychiatrist R. D. Laing illustrated something of the complex, recursive mentality that can arise when social relationships go wrong—or perhaps causes them to go wrong. Here is an excerpt:

JILL: I'm upset you are upset.

JACK: I'm not upset.

JILL: I'm upset that you're not upset that I'm upset that you're upset.

JACK: I'm upset that you're upset that I'm not upset that you're upset that I'm upset, when I'm not.

The evolutionary biologist William D. Hamilton wrote of 'people people' and 'things people'. People people are those who are obsessed with other people, who love gossip, read novels, perhaps think others are thinking about them. Among things people you have computer geeks, engineers, and many scientists—who probably don't give a damn what others think. It's hard to avoid the impression that women tend to be people people, and we unfeeling men tend to be things people—although I can think of many exceptions, including Temple Grandin. Well some, anyway, although as a male I probably wouldn't notice. In any event, we need things people as much as we need people people, since the world is as full of complicated things as of complicated people.

The philosopher Daniel Dennett referred to mind-reading as the 'intentional stance', which means that we tend to treat people as having intentional states. The notion of intentional state is here used rather broadly, and not just the intention to act in a particular way. It includes other subjective states such as beliefs, desires, thoughts, hopes, fears, hang-ups. According to the intentional stance, we interact with people according to what we think is going on in their minds, rather than in terms of their physical attributes—although there is a bit of that too, as I recall from early days on the rugby field, or indeed from days of courtship. When you meet a stranger in a dark alley your reaction may be guided partly by the intentional stance, based perhaps on facial expression, but perhaps also by what might be termed the 'physical stance', based on just how big a hulk the stranger is.

Medical practitioners and even brain surgeons may generally treat people as things, to be fixed mechanically when something goes wrong—a bypass here, a removal of brain tissue there, or a medicine to grapple physically with some internal invader. Psychologists seem to vary. Behaviourists treat animals and humans as objects that simply 'behave', with no mention of the mind. Temple Grandin is a natural behaviourist. Social psychologists are more interested in personality, attitudes and beliefs. Clinical psychologists tend to see psychological problems as mental rather than physical, to be treated with talk rather than drugs. Architects and designers need to be a bit of both, understanding physical as well as aesthetic demands. It's all very well having shoes that look elegant if you can't get your feet comfortably into them.

Just as we may endow people with physical properties, so we sometimes endow physical objects with human-like personalities or subjective states. Perhaps because of their capacity for interior accommodation, cars, ships, airplanes and even houses are often given female characteristics or referred to as 'she'. My father's farm truck was called Lucy, although I once owned a car that answered to the name of Stanley. Throughout history, and perhaps prehistory, people have personified inanimate objects, such as the stars and planets, and have bestowed human properties on non-human animals. People treat their pet cats and dogs as though they were people. Children's stories, in particular, are full of talking animals, from Winnie-the-Pooh to the Big Bad Wolf, from Donald Duck to Little Pig Robinson.

In the quest to identify what might be unique to the human mind, one might well ask whether non-human animals have a theory of mind. In fiction, perhaps, they do. Eeyore, the morose donkey in *Winnie-the-Pooh*, at one point complains: 'A little consideration, a little thought for others, makes all the difference.' In real life, some animals do seem to show empathy toward others in distress. The primatologist Frans de Waal photographed a juvenile chimpanzee placing a consoling arm around an adult chimpanzee in distress after losing a fight, but suggests that monkeys do not do this. However, one study shows that monkeys won't pull a chain to receive food if doing so causes a painful stimulus to be delivered to another monkey, evidently understanding that it will cause distress. Even mice, according to another study, react more intensely to pain if they perceive other mice in pain. It is often claimed that dogs show empathy toward their human owners, whereas cats do not. Cats don't empathise—they exploit.

Understanding what others are thinking, or what they believe, can be complicated, but perceiving emotion in others is much more basic to survival, and no doubt has ancient roots in evolution. Different emotions usually give different outward signs. In Shakespeare's *Henry V*, the King recognises the signs of rage, urging his troops to

> ...imitate the action of the tiger;
> Stiffen the sinews, summon up the blood,
> Disguise fair nature with hard-favour'd rage;
> Then lend the eye a terrible aspect...

The human enemy will read the emotion of Henry's troops, just as the antelope will read the emotion of the marauding tiger. Perhaps the best treatise on the outward signs of emotion is Charles Darwin's *The Expression of the Emotions in Man and Animals*, which details the way fear and anger are expressed in cats and dogs, although he does not neglect the positive emotions:

> Under a transport of Joy or of vivid Pleasure, there is a strong tendency to various purposeless movements, and to the utterance of various sounds. We see this in our young children, in their loud laughter, clapping of hands, and jumping for joy; in the bounding and barking of a dog when going out to walk with his master; and in the frisking of a horse when turned out into an open field.

We might well wonder, though, whether animals do go beyond reading the expressions of emotions, and understand what others are thinking. A good deal of attention has been directed to chimpanzees, our closest non-human relatives. It seems fairly clear that chimps have some understanding of what another chimp can or cannot see. In one study, a chimpanzee approached food when a more dominant chimpanzee could not see it, but was reluctant to do so when the dominant one could see it. Again, a subordinate chimpanzee retrieved hidden food if a dominant chimpanzee was not watching while the food was being hidden, or if the food was moved to another location while the dominant chimp wasn't watching. The subordinate also retrieved food if a dominant chimp watched it being hidden, but was then replaced by another dominant chimp who hadn't watched, suggesting the subordinate could keep track of who knew what.

These are examples of tactical deception. Deception itself is widespread in nature, whether in the camouflage of a butterfly wing or the uncanny ability of the Australian lyrebird to imitate the sounds of other species—including, it is said, the sound of a beer can being opened. Tactical deception, though, is that in which the deception is based on an understanding of what the deceived animal is actually thinking, or what it can see. Two psychologists from St Andrews University in Scotland, Andrew Whiten and Richard Byrne, once put out a general call to researchers studying primates in field settings for anecdotes demonstrating tactical deception. They screened the reports to rule out cases in which the animals might have learned to deceive through trial and error, and concluded that only the four species of ape occasionally deceived on the basis of an understanding of what the deceived animal could see or know. Even so, there were relatively few instances. Chimpanzees were alone in meeting nine of thirteen different classes of deception, whereas gorillas met only two. Perhaps our primate cousins are exceptionally cooperative and trusting, or their capacity for theory of mind is limited compared with the human predilection for deception, from petty lies to downright fraud.

In 1978, psychologists David Premack and Guy Woodruff wrote a classic article with the beguiling title 'Does the chimpanzee have a theory of mind?' This led to a good deal of research, but the issue is still not entirely settled—it seems that we humans, even the experts, are quite good at reading the minds of other humans, but not so good at reading the mind of the chimpanzee. Nevertheless, two of the experts in the field, Josep Call and Michael Tomasello, concluded that 30 years of research had shown chimpanzees to have an understanding of the goals, intentions, perceptions and knowledge of other individuals, but no understanding of their beliefs or desires. No one has yet convincingly shown that chimpanzees can attribute a false belief to another chimp.

However, the champions of mind-reading among animals are probably not chimpanzees, but our best friends. Dogs seem to have an almost uncanny knack of understanding what's going on in the minds of human beings. They readily understand pointing. For instance, if two containers are placed in front of a dog, and a person points to the one that contains food, a dog will understand that the gesture of pointing is designed to indicate the food. The food is hidden from the dog's view, and experiments show that the choice is not based on smell. They will also go for the food if the person points to a container behind them. They will even choose the correct container if a person simply marks it by placing some object on top of it. Puppies without much human experience act in the same way. Chimpanzees, in contrast, are much worse at such tasks.

Dogs are descended from wolves, and wolves do not respond in the same way. The key to mind-reading in dogs is domestication. Surprisingly, though, the domestication of dogs seems not to have been driven by humans, at least initially. Brian Hare, who despite his name likes to be known as the 'dog guy', suggests that dogs evolved from packs of wolves who scavenged from the rubbish left by humans, and those most likely to survive were those who were least afraid of human contact, and became comfortable in the presence of humans. In Hare's words, it was 'survival of the friendliest'. At some point, though, humans seem to have capitalised on the friendliness of the dogs and embarked on further selective breeding, to produce the extraordinary variety of dogs we see today.

(My favourite? The Nova Scotia duck tolling retriever,[*] bred to wag its tail to attract ducks, luring them close to hunters.) Some of them have been bred back to a condition of unfriendliness, serving as guards ready to attack intruders. 'Cave canem', as they used to warn in Rome. Occasionally, we read of savage attacks on people by dogs, often followed by threats to have such breeds exterminated, but most dogs are wonderfully friendly and faithful, and experts at reading the human mind.

Another species to have become domesticated independently of human influence is the bonobo, close cousin of the chimpanzee, and sharing with the chimp the honour of being our closest non-human relative. In personality, though, chimps and bonobos are opposite. Chimpanzees are aggressive and competitive, with males often attacking females and the young, while bonobos are friendly, caring and sharing, and use sex rather than fighting to resolve conflicts. Sadly, they were almost exterminated in the Congo Basin in the bushmeat trade, until a sanctuary called Lola ya Bonobo, meaning Paradise of the Bonobos, was established. Curiously, increased domestication seems to be accompanied by a decrease in brain size. Dogs have smaller brains than wolves of equivalent body size, and bonobos have smaller brains than chimps. And we humans have slightly smaller brains than the now extinct Neanderthals, our closest non-living relatives. So beware of men with big heads, and take comfort from Oliver Goldsmith's poem 'The Village Schoolmaster':

---

[*] Well, I've never actually met one. It's the name that I really like.

And still they gazed, and still the wonder grew,
That one small head could carry all he knew.

The fundamental question that haunts research on the animal mind is whether there is a discontinuity between ourselves and other species. Much of religious teaching is built on the premise that we humans are indeed on a different plane, closer to angels than apes, even if our sins have caused us to fall a little lower. René Descartes, too, argued that humans are unique through the possession of a non-material mind, whereas animals are mere machines. Opposed to that, of course, is the Darwinian mantra I quoted in the previous chapter: 'The difference in mind between man and the higher animals, great as it is, certainly is one of degree and not of kind.' My guess is that there is indeed a continuity between ourselves and other primates, at least in our ability to understand what's going on in the minds of others, but with greater complexity in our own species.[*] Indeed, as I illustrated earlier, we humans seem able to take that ability to depths of recursion well beyond that evident in chimpanzee society. This may well have been driven by increasing loops of deception, the product of the so-called Machiavellian mind—as Machiavelli himself put it in *The Prince*: 'It is double pleasure to deceive the deceiver.' The spiral of deception and intrigue has been cynically described by Nicholas

---

[*] In an article published in 1997, Thomas Suddendorf and I proposed that theory of mind drew on the same mechanisms as mental time travel, and argued that both were unique to humans. My own view now is that there is greater continuity between species than I thought at the time. For a persisting view that humans are indeed unique in these respects, see Suddendorf's excellent new book *The Gap: The Science of What Separates Us from Other Animals* (New York: Basic Books, 2013).

Humphrey as a 'self-winding watch to increase the general intel-lectual standing of the species'.

Whether to deceive or inform, we humans seem to delight in making mental journeys into the minds of others, and indeed create fictional characters for the purpose. Young children, espe-cially in the preschool years, often create imaginary companions, invisible friends with whom they share confidences. Together with the ability to travel mentally in time, travelling mentally into the minds of others provides the platform for one characteristic that does seem to be distinctively and universally human—storytelling.

6.
STORIES

• • •

My voice goes after what my eyes cannot reach,
With the twirl of my tongue I encompass worlds
and volumes of worlds.
　　—Walt Whitman, from 'Song of Myself'

We humans may not be entirely unique in taking mental journeys through time and space, and into the minds of others. Rats may well have a limited capacity for mental time travel, imagining past or future activity in a maze, and chimps may have some inkling of what's in another's mind. We may like to think that our own mind-wanderings are richer, more interesting, more personally invasive than those of other species, but what does seem to be special about humans is the capacity to share our mind-wanderings. We tell stories, taking others with us on our meanderings. The literary scholar John Niles suggested that our species should be renamed *Homo narrans*—the storytellers.

Although we seem to be the only species that tells stories, our capacity for narrative probably has evolutionary precedents. The writer and literary theorist Brian Boyd, among others, suggests that stories derive from play, an activity that goes far back in evolution. Playing means doing something for recreation or enjoyment rather than for a serious purpose, and often entails pretending to be something that one is not. Many animals play, from frolicking kittens to perky parrots, from bounding puppies to cheeky monkeys. I have read that reptiles have their playful moments. Social species play more than solitary ones do, and species that hunt play more than those that are hunted, perhaps because hunting requires more ingenuity than escaping does, and play is a way of trying out new strategies. It often takes the form of mock chasing or attacking, and serves to increase survival fitness by providing practice for the real thing. When the puppy makes a play bite, it is understood by the biter and the bitten that it is not for real. No blood is spilled, no flesh removed. Dogs announce the desire to play with a characteristic 'play bow', crouching on the forelimbs while leaving their hind limbs straight, and wagging their tails. My four-year-old granddaughter is more direct—she simply says 'play with me'.

Play also takes place between species. People and dogs, in particular, seem to love to play together. The most common game is 'fetch', in which we throw sticks, balls or Frisbees for the dog to fetch, and this is often followed by the dog playfully refusing to let go after returning the object to the thrower. Jay Mechling quotes a report of a man playing a game of 'banana cannon' with his dog Shana:

> Every morning at breakfast time when John peels his banana,
> Shana gets excited. She sits on the floor, approximately five feet
> away from John, and waits for John to play 'banana cannon'. John:
> 'I take a piece of banana and shoot it like a cannon out of my mouth.
> She's real good. Gets it from way back.'

It's not just people and their dogs. The anthropologist Gregory Bateson tells of a game played by a tame female gibbon and a tame female puppy. The gibbon would come down from the rafters of the porch and lightly attack the puppy, who would give chase. The gibbon would not retreat to the safety of the rafters, but would run down the corridor and into the bedroom. Since she would be cornered there, the game would reverse, and she would now chase the puppy back to the porch. She would then retreat to the rafters, and the whole game would start all over again. The sequence would sometimes be repeated seven or eight times.

But such games are not stories, because they are played in the present. One additional ingredient, then, is the 'once upon a time' element that takes the action away from the here and now and into the past or future, or into other places, or into the lives of other people, real or imaginary. A second ingredient is narrative. Stories have a complex, usually sequential, structure—events unfold through time, often in elaborate ways. And the third ingredient is that stories are shared beyond those that feature in them. Personal mental time travels become shared mental time travels, whether in the tales told of travels abroad or of imaginary adventures, or indeed of what one is planning to do on the next trip. Stories are a mixture of actual experience and made-up fantasy, of work and play. It is through stories that experiences of individuals become the experiences of a social group, or even an entire culture.

The emergence of stories from play is evident in the lives of preschool children, who seem to live in a world of make-believe. Indeed, before they go to real school, many young children are sent to playschool, whose very name speaks of the importance of play in their young lives. They begin to experiment with simple make-believe stories around age two or three, and tell competent stories by age five or six. They seem especially to relish stories with elements of danger, as though playful exposure to fearful events might help them cope with true danger later in life. The world of the three-year-old is one in which fact and fantasy are blended, and there is the fear that the Big Bad Wolf, or some other fearsome creature, is really lurking behind a tree—or worse, under the bed. Children's nursery rhymes and fairy stories, from 'Jack and Jill' to 'Little Red Riding Hood', tell of frightening events and disasters, usually (but not always) narrowly averted.

Stories probably originated in our hunter-gatherer past, as our early forebears relayed their foraging experiences. Some sense of this can be gleaned from present-day hunter-gatherers. Among the Aché people of eastern Paraguay, each man is said to report to the others in detail on every game item that he encountered that day, and the outcome of the encounter. This enables the group to become familiar with the terrain, likely locations of game, hunting techniques, successes and failures. Foraging as a way of life, involving exploration over wide terrain, probably goes back to the early Pleistocene, beginning some 2.6 million years ago and extending to around 12,000 years ago. During this era, as foraging gradually expanded to include active hunting, and as the range and diversity of terrain increased, the pressure to communicate effectively would also have increased. Children, too, would be captivated by the tales told by the men, and perhaps repeated by the women, and so would gain knowledge about food sources and hunting techniques before themselves going on the hunt.

Grandparents can be useful too. In the Jicarilla Apache of northern New Mexico, as in other present-day hunter-gatherer societies, it is the elders in the extended family who tell the stories to the children. This arrangement may well have originated in the Pleistocene, and might help explain why humans have evolved to live well beyond child-bearing years to become the bearers of wisdom and stories for their grandchildren. Michelle Sugiyama also suggests that the telling of stories helps explain the prolonged juvenile phase—kids have a lot to learn before they're fit for adult life and the rigours of the hunt, or indeed of child-rearing itself.

Hunting and gathering are risky, and were especially so on the African savannah during the Pleistocene, when dangerous sabre-toothed cats roamed the plains. This gave added advantage to sharing knowledge and expertise, not least so that the wisdom of those killed in action would not die with them. But stories go well beyond the communication of knowledge, incorporating the sense of play and fantasy, the invention of imaginary places and imaginary minds, the creation of cultural beliefs. In many ways, then, the ability to tell stories enhanced the survival of the group, even if sometimes at the expense of the individual.

Storytelling also established social hierarchies. In traditional societies, at least, this seemed to apply particularly to men, in whom the ability to hold forth in public was the avenue to status and influence. For New Zealand Māori, writes Anne Salmond, 'oratory is the prime qualification for entry into the power game'. Males are the loudmouths, with the deepening of the voice seemingly designed to command attention: Salmond goes on to write that a great Māori orator 'jumps to his feet with a loud call and immediately dominates the speaking-ground'. The social anthropologist David Turton similarly wrote of the Mursi in south-western Ethiopia that 'the most frequently mentioned attribute of an influential man is his ability to speak well in public'. The same may be true even in city life. In the inner-city neighbourhoods of Philadelphia, according to the anthropologist and folklorist Roger Abrahams, the African-American 'man of words, the good talker, has an important place in the social structure of the group, not only in adolescence but throughout most of his life'.

As I write, one man has succeeded another as the Prime Minister of Australia, and three male aspirants are vying to lead the opposition party in New Zealand to compete with the current Prime Minister, also a male. Vocal eloquence seems to feature prominently in these bids for power. One hopes, of course, that eloquent orators also have something to say.

But it's not all male, since until quite recently the Prime Ministers of Australia and New Zealand were both female, and in modern society, at least, women are if anything superior to men in the use of language. The way in which language is used, though, may differ. Men seem more inclined to use language as a form of public display, like peacock feathers, whereas women are more likely to engage in intimate talk, to gossip, to use language to seek companionship rather than power. Women's talk, perhaps, is more subversive, a way of communicating that carries subtleties undetected by us blustering males. Or so I have been led to believe.

And so language itself was born

In the early stages, perhaps, stories were told as pantomimes, as people acted out their experiences. But pantomime is inefficient and often ambiguous, and needed to be developed into a system of symbols whose meanings were clear, and understood by members of the community. Once, in a hotel in Moscow, I tried to ask at the front desk for a corkscrew, and having no knowledge of Russian pantomimed the act of opening a bottle, pouring imaginary liquid into an imaginary glass, lifting it to my lip, and making glugging noises. This caused consternation behind the desk, until they understood what I wanted. Consternation turned to hilarity and they found me the desired object. It would have been much more efficient if I had been able simply to ask for a corkscrew.

In the early Pleistocene, then, the complex activities undertaken by our forebears, such as hunting down and killing an animal, may have initially been acted out bodily, but then 'conventionalised' to make the meaning clear. Instead of representing the action in a holistic, visual fashion, separate actions might be developed to refer to the animal, a spear, and act of throwing, the location, and perhaps the time (yesterday, this morning). Each act could be reduced to a standard form, and need no longer retain the pictorial element of pantomime. Communities could come to agree on the meanings of individual acts, and pass them on to the children. This process can be seen in gestural form in the development of sign languages invented by deaf communities.

Once gestures were conventionalised, the element of pantomime disappeared, and vocal gestures largely replaced manual ones. Even so, most of us, and especially Italians, gesture with our hands as we speak, and this often provides pictorial or spatial cues to elaborate what we're trying to say. Sometimes we resort to pure pantomime. Try asking people to tell you what a *spiral* is. Words generally fail them, and they resort to pantomime. Sign languages do retain an element of pantomime, but skilled signers do not notice the pantomimic element. Gestures have become symbols, not moving pictures. Whether gestured or spoken, these conventionalised symbols are called words.

Rules could then be established to convey sequences and relations between story elements. These rules might dictate the order in which the words should occur. They too can be arbitrary, but once established are necessary to make narrative meaning clear. So it is that grammar was born. Many simple events involve what linguists call an actor, an action and a patient; for instance, the event might have been that a woman (actor) picked (action) an apple (patient). These constituents are represented in word form as subject, verb and object, and the order in which they are produced is entirely a matter of convention. English is an SVO language, but the majority of languages, like Latin, are SOV, placing the verb last. All six possible orders are to be found among the world's 7000 or so languages. The rarest are OSV languages, of which only four are known (in case you are travelling, they are Warao in Venezuela, Nadëb in Brazil, Wik Ngathana in north-eastern Australia, and Tobati in West Papua, New Guinea).

Some languages use other devices to mark the differences between subject, verb and object, and indeed between the many other kinds of symbols to specify place, time, quantity, quality, and other details needed to set a scene or an event in words. Some such languages are called scrambling languages, because the order no longer matters. Walpiri, an Australian language, is an example, and Latin can be scrambled quite a bit without altering meaning because the complex system of suffixes makes it clear which is the subject and which the object, as well as specifying tense, number, and the like. But however it is structured, language is the device that allows us to tell stories of a complexity limited only by powers of memory and ability to sustain attention, located at times and places away from the present, and sometimes venturing into the minds of others.

If language indeed grew from pantomime, as many have conjectured, its earlier origins must have been gestural rather than vocal, and in this respect may go a long way back into our primate heritage. The mechanics of language may derive not from vocal calls, but from the use of the hands for grasping. The vocal calls of monkeys and apes are very largely emotional and instinctive, tied to the immediate situation, and largely useless for storytelling. The hands, in contrast, are used in a flexible, intentional way, and seem almost custom-designed for conveying information about events. Indeed, the notion of grasping still seems embedded, if only metaphorically, in our very speech. The word *grasp* is itself often used to mean 'understand', if you grasp my meaning. *Comprehend* and *apprehend* derive from Latin *prehendere*, 'to grasp'; *intend*, *contend* and *pretend* derive from Latin *tendere*, 'to reach with the hand'; we may *press* a point, and *expression* and *impression* also suggest pressing. We *hold* conversations, *point* things out, *seize upon* ideas, *grope for* words—if you *catch* my drift. It works visually, too, as when you *see* what I mean, as I hope you do.

The invention of speech must have occurred well after the line leading to humans split from that leading to the great apes. Attempts to teach apes to talk have failed rather dismally, but chimpanzees, bonobos and gorillas have become quite proficient at learning simplified forms of sign language. The star is the bonobo Kanzi, reared by Sue Savage-Rumbaugh; he communicates by pointing to symbols on a specially designed keyboard, and supplements these gestures with signs that he apparently picked up by watching the sign language used by Koko, a signing gorilla. Apes in the wild make wide use of bodily actions to communicate with one another, often in the context of play. Robin Dunbar has suggested that the origins of language lie in grooming, a gentle activity of picking and cleaning the fur of another animal, and one that is important in cementing social relationships. A related act of communication is the 'directed scratch', in which a chimpanzee scratches the part of its body where it wants to be groomed by another.

Just when hand gestures were replaced by spoken words is a matter of conjecture. Jean Auel's novel *The Clan of the Cave Bear* is set in the Ice Age, 27,000 or so years ago, when early humans and Neanderthals co-existed. A five-year-old girl, Ayla, is orphaned after an earthquake kills the rest of her family, and is eventually adopted into a Neanderthal community. The Neanderthals in the story cannot speak, and communicate in sign language. I should not, of course, take a fictional novel to be acceptable scientific evidence, tempting as it is to do so, but Auel is nevertheless something of an expert on early humans and Neanderthals, and the use of sign language by Neanderthals is a theme in her other novels as well. Curiously, though, the Neanderthals in *The Clan of the Cave Bear* not only could not speak, but they also could not laugh or cry, and when they saw Ayla weeping they thought she had an eye disease. Even chimps can laugh. In Auel's novels, the Neanderthals were also able to communicate through telepathy.

The real Neanderthals did interbreed to some degree with our own species. They died out only some 30,000 years ago, and I suspect that, like us, they were capable of articulate speech. The shift from manual to vocal communication probably occurred gradually during the Pleistocene, and is still not complete. We all gesture manually as we speak, and the sign languages used by the deaf and some other communities are as effective and linguistically sophisticated as speech. Why, then, would spoken words have been introduced, and become so dominant? I think there are many answers. Except for our continued disposition to wave our arms about as we speak, speech frees the hands for other things, such as using and making tools, carrying things, and tending to the needs of infants. Speech is itself a form of gesturing, involving movements of the tongue, lips and vocal cords, but is located tidily within the mouth. It is an early example of miniaturisation, and interferes only with intermittent activities like eating and kissing. We tell our children not to speak with their mouths full, and sympathise with the poet John Donne's agonised cry from his 1633 poem 'The Canonization': 'For God's sake hold your tongue, and let me love.'

Speaking is also much less tiring than gesturing manually, since it involves much smaller movements and piggybacks on breathing, which we have to do anyway in order to survive. Speech works at night, or when speaker and audience are not in visual contact—a property exploited by radio and cellphone. I could go on and on.[*]

But whether speaking or signing, we humans gained a profound skill unapproached by any other species. Our closest non-human relatives, the great apes, do not tell stories, even when they gesture. At most, they make simple requests, or respond to simple instructions. The capacity to relay narratives to an audience, through the invention of grammatical language, does seem to be a distinctive characteristic of humans. Whether the distinctiveness lies in the internal construction of imagined events, or simply in the telling of them, remains something of a moot point. Either way, one might

[*]   And indeed have done in my 2002 book *From Hand to Mouth: The Origins of Language* (Princeton: Princeton University Press), which has since appeared in paperback (2003) and been translated into Turkish (2003), Italian (2008) and Japanese (2009).

well agree with the French psychologist and psychotherapist Pierre Janet, who wrote: 'narration created humanity'.

The stories we tell

Through the Pleistocene, then, our forebears evolved the characteristics that we think of as human. It was the Pleistocene that saw the emergence of the genus *Homo*, of which *Homo sapiens* is the only remaining species. The brain tripled in size, and the fully upright stance and striding gait enhanced the ability to wander over wide terrain—no doubt contributing to the wandering of the mind as well as of the body. Our forebears established what has been called the 'cognitive niche', surviving the dangerous environment of the African savannah through the sharing of knowledge and the telling of stories. Stories bind peoples together, and create culture. Every culture seems to have its tales of heroism and discovery, establishing a sense of common ancestry. In modern times this is largely conveyed through the written word, but in preliterate societies stories were told down the generations through spoken words or gestures. Many of these are perhaps still locked in languages inaccessible to outsiders, but those that are known have many features in common.

We have no record of stories that predate the arrival of our own species some 200,000 years ago, but stories handed down through the generations of present-day cultures can provide some appreciation of their nature and content. They seem to have as much to do with establishing myths and creation stories as with the sharing of practical knowledge. The indigenous Australians have told stories that may well go back at least 50,000 years to their arrival in Australia, and shortly after the exodus from Africa. They tell of the Dreamtime, a sacred era when ancestral spirits created the world. Some of these god-like figures are more powerful than others. In south-eastern Australia, it was Biame, the All-Father, who first created the animals and then used them as models to create humans; in the Northern Territory, it was the sky god Altjira of the Arrernte people, who created the earth. Dreamtime persists in the Dreaming, the continuing set of beliefs and traditions. Dreaming stories were carried by culture heroes, and expressed in songs and dances across Australia, even through different language groups. They cover a great many topics, about people, places, laws and customs. Children exist in the spirit-child before being brought to life through birth to a mother, and persist eternally after life. Christian tradition similarly tells stories of creation, all-powerful gods, and eternal life after death.

New Zealand Māori have a much more recent history in their adopted country, having arrived in New Zealand only around 750 years ago, but they too have retained intricate stories told by word of mouth. Māori legend goes back to the demigod Māui, who had magical powers and lived in a place called Hawaiki.* One day, out at sea, he dropped his magic fishhook over the side of the boat, and felt a powerful tug on the line. With the help of his brothers, he pulled up a large fish, which they called Te Ika a Māui (Māui's fish), and which became the North Island of New Zealand. The South Island of New Zealand was Māui's boat, called Te Waka a Māui, and Stewart

---

* Not Hawaii, as some have guessed. Māori settlers probably sailed from somewhere in central-east Polynesia.

Island, the smaller island at the southern end of the country, was Māui's anchor, Te Punga a Māui, which held the boat steady while Māui reeled in the giant fish. Although Māui caused these events, it was the great Polynesia navigator Kupe who discovered the new land, Aotearoa—'Land of the Long White Cloud', otherwise more prosaically known as New Zealand. Of course, there is also much more to Māori lore, which includes accounts of the creation of the world, stories of battles, songs, poems and prayers—elements again to be found in virtually all religions.

One curious exception is the Pirahã, a remote tribe of people on the Amazon in Brazil. Daniel Everett went there as a missionary with the intent of studying their language so he could translate the Bible for them. He discovered that their language was impoverished by western standards, with a small vocabulary and only indirect ways of referring to the past or future. According to Everett, they don't create fiction and have no creation stories or myths. The Pirahã language, though, is related to another language called Mura, which evidently does include rich texts about the past. One possibility, then, is that the Pirahã people separated from the Mura at some stage, and in the process lost a sense of their historical past, and indeed seem to repress the personal past as well. Everett spent several years living among them, and records that they are in no way intellectually impoverished—they were quite happy to discuss cosmology and notions about the origins of the universe with him, in spite of having no material of their own to draw on. Indeed, he seems to have been so impressed with their way of life that he was converted from Christianity to become an atheist, and is now a professor of linguistics in the United States.

Most societies, though, do have stories and creation myths, and in preliterate cultures these are typically expressed in verse or song. Rhyme and metre seem to be strong aids to memory. With the invention of writing, there is no longer the strong need for such devices, although children are still taught rhymes to help them remember lists, such as the alphabet, the elements of the periodic table, or the colours of the rainbow. Then there is the poem that gives the first 21 digits of the mathematical constant *pi*:

Pie
I wish I could determine *pi*
Eureka, cried the great inventor
Christmas pudding, Christmas pie
Is the problem's very centre.

You simply count the number of letters in each word, and put the decimal point after 'Pie' (3.14159 26535 89793 23846). However, a downside of using jingles to remember things may be the earworms that I mentioned in Chapter 1—the songs and ditties that take over the mind and won't go away. If the 'Pie' jingle won't go away, I suggest dumping it on someone who might actually use it. It's probably not a very useful gift, though; if you really want to memorise *pi* to lots of decimal places, you're better off using the method of loci, as illustrated in Chapter 2, or finding it through Google.

Even after the invention of writing, and later the printing press, epic tales persisted for a while in the form of long poems, whose rhyme and metre remained as an aid to memory, helping ensure faithful transmission from one generation to the next. The earliest known story in literature may be the *Epic of Gilgamesh*, which goes back some 4000 years. Gilgamesh, a Sumerian king, befriends Enkidu, a wild man created by the gods to divert Gilgamesh from oppressing his people. Gilgamesh and Enkidu journey to Cedar Mountain to defeat Humbaba, guardian of the mountain. They then defeat the Bull of Heaven, sent by the goddess Ishtar to exact revenge on Gilgamesh for spurning her advances. In revenge, the gods then kill Enkidu. In distress, Gilgamesh goes on a long journey to seek immortality. He dies, but his fame lives on for his great accomplishments, and the story itself has provided the basis for many later works of fiction. Such stories carry a full range of emotions, and establish heroes and villains that act as models for the way people behave in society.

Other examples are the *Iliad* and *Odyssey*, two famous Greek poems written by Homer, dating from around the eighth century BC. More recent examples include Dante's *Inferno* and Chaucer's *Canterbury Tales* from the fourteenth century, John Milton's *Paradise Lost* from the seventeenth century, Samuel Taylor Coleridge's *Rime of the Ancient Mariner* from the late eighteenth century, or Lord Byron's *Don Juan* from the nineteenth century. Epic poems have been well surpassed by stories told in prose or long-running television soaps, although the form is continued by the Australian author and poet Clive James in his satirical verse epic *Peregrine Prykke's Pilgrimage Through the London Literary World: A Tragedy in Heroic Couplets* which was published in 1974. I'm told he's working on another.

Brian Boyd points out that religious ideas themselves owe less to doctrine than to stories, and religious stories are typically stories of magical deeds. In the Bible, Psalm 77:14 proclaims: 'You are the God who performs miracles; you display your power among the peoples.' The four gospels of the New Testament record 37 miracles performed by Jesus, including healing the sick, turning water into wine, and walking on water, and Jesus himself was held to be the son of God, born of a virgin mother. The Qur'an, the holy book of Islam, is itself regarded as a miracle, revealed to the prophet Muhammad from Allah through the archangel Gabriel.

Crime fiction

Once writing was invented, stories became much more varied and widely dispersed. Even so, they still play a big role in establishing heroes, and reinforcing moral values. This can be illustrated from a perhaps unlikely source: crime fiction. Murder and other crimes have always featured in stories, from the Bible to Shakespeare, although in modern crime stories they have taken on new conventions, built largely on the trappings of industrial society. You might think the human obsession with murder would only encourage mayhem rather than peaceful cooperation in society, but crime stories, like the epics of old, are really morality tales, since the perpetrator is always caught and duly punished. Crime stories, as we understand them, are largely a phenomenon of western culture, and are in many respects peculiarly English, but they express universal themes.

What perhaps distinguishes modern crime stories from older tales of murder and brutality is the emergence of the detective as the hero. Detective stories as a popular genre go back only to the mid-nineteenth century, with authors such as Edgar Allan Poe and

Wilkie Collins. The archetypal detective-hero is Arthur Conan Doyle's Sherlock Holmes, who was not only a moral guardian in his pursuit of the dastardly Moriarty, but also a clever geek who used extraordinary powers of observation and deduction to solve crimes. He was even a model for aspiring scientists—catching the criminal is a bit like catching the Higgs boson (although less expensive). Of course, Holmes did not really exist outside of Conan Doyle's imagination, but he became so popular that he was widely regarded as a real person, if not a demigod. When Conan Doyle killed him off in 'The Final Problem', published in 1893, public pressure was such that he brought him back to life in 'The Adventure of the Empty House', set in 1894 but not published until 1903.

Sherlock Holmes established a tradition of fictional detectives with different, often exotic, identities, luring us into their minds, and in so doing expanding the way we see ourselves. John Buchan's Richard Hannay, hero of *The Thirty-Nine Steps* and other spy stories, is perhaps the archetype of the values of the English public school, stiff upper-lipped and fearless—or at least conditioned not to show fear. In a similar tradition are Bulldog Drummond from the novels of H. C. McNeile (who wrote under the pen name of 'Sapper') and Ian Fleming's James Bond, who continues to create mayhem in popular movies. But perhaps the reading public grew weary of jingoistic heroes, and fictional detectives have generally taken on more gentle and often eccentric personae. Dorothy Sayers' novels featured Lord Peter Death Bredon Wimsey as the archetypal English aristocrat with too many names. Agatha Christie created the fastidious Belgian Hercule Poirot, and when she tired of him introduced the elderly spinster Miss Jane Marple, who seemed able to combine knitting with crucial observation of giveaway clues. G. K. Chesterton's sleuthing priest Father Brown seems to be enjoying a revival on our television screens. More recent examples are Henning Mankell's morose Kurt Wallander, Ian Rankin's dissolute John Rebus, and Sara Paretsky's tough-minded female V. I. Warshawski.[*] Real-life detectives, it has to be said, are a much more prosaic lot, at least if we can judge from their occasional appearances on television, or before real-world suspects gathered in the living room. (Does this actually happen?)

Entering the mind of a fictional detective may also allow us into places or elements of society from which we may normally be barred. In a recent interview, the Scottish crime writer Ian Rankin had this to say:

> A detective is the perfect character, the perfect means, of looking at society as a whole. I can't think of any other character you could use that allows you access to any area of society.... [The detective allows] access to the banks, the politicians, the CEOs, the people who run business, but also the dispossessed, the disenfranchised, the unemployed, the drug addicts, the prostitutes.[†]

Crime fiction, then, is truly an exercise in guided mind-wandering, escorting us into different places, different times, different minds.

Another feature of crime stories is that they alert us to dangerous events that might happen (but we hope they won't), and so provide scenarios that might follow if they do, and make us better prepared to deal with them. Then there's the dark side. By exposing

---

[*] Her initials stand for the more lady-like 'Victoria Iphigenia'.
[†] From an interview with the *New Zealand Listener* of 3 November 2012.

the mistakes that lead to the criminal being caught, crime stories may help the readers themselves get away with murder. But it can go the other way, with fact pre-empting fiction. The 1994 movie *Heavenly Creatures*, directed by Peter Jackson, is based on the true story of two New Zealand schoolgirls who murdered the mother of one of them. They were, of course, caught and imprisoned, and one is now an internationally known author of crime fiction, writing under the name of Anne Perry.[*]

Beyond murder

Of course, not all fiction is murderous. Many novels depict everyday life, but with an imaginative overlay that enhances understanding or emotional involvement. They create characters that our own minds can wander into, leading us into vicarious adventures and crises. Plays and novels also serve as social commentary. The novels of Charles Dickens not only provide a vivid portrayal of nineteenth-century London, but were also crafted to highlight the conditions of the poor, and bring about social reform. Dickens pioneered the serialisation of novels, so that readers would eagerly await each next instalment—a technique that persisted in serialised radio productions and more recently in television series. He also perfected the art of literary caricature, creating such memorable but exaggerated characters as Fagin, Uriah Heep and Mr Pickwick.

Just like the gods of ancient stories, the characters of modern fiction often extend beyond caricature to characters that transcend

---

[*] She has herself acknowledged her past, in public interviews.

normal human capabilities. Children's stories, in particular, are alive with talking animals, fairies, magicians and other supernatural beings, well illustrated by the extraordinary success of the *Harry Potter* series. Is the supernatural adaptive? Perhaps the overstretching of the imagination allows us to better understand what might be possible, although it is perhaps more often a product of wish-fulfilment. If we could fly, become immensely strong, control events with our thoughts—we could overcome many of our problems in coping with the world. James Bond and Superman belong to a long tradition of heroes with superhuman qualities.

Fiction, like other forms of play, is sometimes dismissed as mere fantasy, an escape from the realities of life. We should discourage our kids from reading comics or watching TV cartoons, some say, and have them help with the dishes or tidy their rooms for once. Studies show, though, that fiction increases empathy and improves mind-reading, making us better able to understand others. Brain-imaging has shown overlap between areas of the brain activated by reading narrative stories and those involved in theory of mind. One study measured the amount of fiction and non-fiction that people read, and found that empathy was correlated positively with the amount of fiction read, but negatively with the amount of non-fiction. Another recent study carried the headline 'Reading Literary Fiction Improves Theory of Mind'. Better a bookworm than a technology nerd if you want to get on in the social world—but then again, as I noted in the previous chapter, we also need people who can fix our washing machines and set up our computers. Donald Hebb, the esteemed Canadian psychologist and neuroscientist, and one of my mentors, used to tell us as graduate students that we could learn more practical psychology by reading novels than from poring over the journals of experimental psychology. More fun, too.

Of course, language is not just used for sharing stories. We also use it to share knowledge, although I've found that the occasional story thrown into a lecture tends to keep the students awake. And knowledge itself is often story-like. Modern physics, for example, is full of eccentric and all-powerful entities, like mesons and baryons and quarks—and of course the God particle itself, the Higgs boson. These are modern equivalents, perhaps, of the demons and gods of ancient mythology.

If there's anything that defines our species as unique, then, it is the telling of stories, and the invention of language as the means of doing so. As I suggested in the previous chapter, other animals, even rats, may well undertake limited mental travels through limited domains, but stories allow us to expand our mental lives to unlimited horizons. Through the power of stories, we learned the means of constructing vast cities and machines, assemblages as multi-storied as they are multi-storeyed. Language itself expanded from the telling of tales to the invention of mathematics, the vast power of computing, the symbolic resources of the internet, the ubiquitous cellphone. Stories combine narrative and play, allowing us to construct edifices, both real and imaginary. Our mental travels took us to the moon or to a landing on Mars long before we could accomplish these journeys by physical means.

But the creative component comes not just from our memories and our playfulness. It also has sources that are detached from our conscious control. These sources are the topics of the next two chapters.

7.
TIGERS IN
THE NIGHT

· · ·

We are such stuff
As dreams are made on, and our little life
Is rounded with a sleep.
—William Shakespeare, from *The Tempest*

The dreams that we have when we're asleep are especially ubiquitous wanderings of the mind. Just as Mitty-like daydreaming does, night-time dreams activate the default-mode network—that widespread mesh of connections in the brain, described in Chapter 1, which lights up when our attention is not focused. Dreams, moreover, are stories, with a narrative structure that unfolds sequentially through time. We live them as though they are real and actually happening, but in this respect they are not like our waking wanderings. Even Walter Mitty would have known, had he himself been real, that he was not actually in a huge, hurtling eight-engined Navy hydroplane, but was driving along a highway beside his rather agitated wife. He was daydreaming, but this is not the same as the dreams that visit us with unfailing regularity when we're asleep.

Although the great majority of dreams are experienced as though real, we occasionally have what are called 'lucid dreams', in which we are aware that we are dreaming. If the dream is unpleasant or frightening, we can then perhaps escape by somehow forcing ourselves to wake up. One technique that seems to work for me is to force my eyes open—a strangely paradoxical strategy since I dream as though my eyes were open anyway. In one recent attempt to do this, though, I somehow propelled myself not into wakefulness, but into another dream that I was awake. Or perhaps I dreamed this whole story. In the words of the American singer Beyoncé, it could even be that 'life is but a dream'.* How would we know?

While we sleep, we fluctuate between periods of rapid eye movement (REM) and non-rapid eye movement (NREM). Dreams are most vivid and sustained during REM sleep, which occurs about every 90 minutes, so we have three to four REM episodes per night. When people are wakened during REM sleep they report that they were dreaming about 80 per cent of the time, but wakening during NREM sleep produces dream reports less than 10 per cent of the time. Curiously, though, people do report that thoughts were going through their minds prior to being awakened from NREM sleep—estimates here range from 23 per cent to nearly 80 per cent. This suggests that thoughts during NREM sleep often have a contemplative rather than a dream-like quality. In the stage of NREM sleep at sleep onset, though, people report brief but vivid hallucinatory experiences, known as 'hypnagogic hallucinations', around 80 or 90 per cent of the time. These are unlike dreams during REM sleep in that they are brief and often static, and do not include the actual dreamer. In REM dreams we are normally and sometimes painfully present as participants.

Parts of the cortex of the brain normally activated by vision itself are also activated by these hypnagogic experiences. A team of Japanese researchers identified patterns of activity in the visual areas of the brain elicited by objects and scenes in the visual world, and then recorded activity in these areas in three volunteers while they were in the onset phase of sleep. The sleepers were then woken

---

* Older readers may recognise the line from the song 'Row, Row, Row Your Boat', first published in 1852. But is anyone *that* old?

up, and asked to describe any visual experiences they were having prior to being awoken. The researchers were able to predict the visual content from the patterns of activity with some 60 per cent accuracy—not perfect, but much higher than expected by chance. With improved imaging technology, we may one day be able to tell exactly what people are dreaming about without having to ask them. That could be the ultimate invasion of privacy.

Dreams very seldom replay past episodes, but are typically made up of fragments of memory, sometimes combined in bizarre ways. Dreamers readily accept impossible events, such as flying, or the face of one person appearing on the body of another. Scenes can switch without cause or reason—one moment I am back in my school dormitory, then suddenly trying to negotiate a dangerous path on a cliff face. Both are based on past events, somehow seamlessly blended in the dream. Although dreams are built from memories, memory for dreams themselves is poor. In fact, virtually all dreams are forgotten, unless we wake up while having them—and even then it's probably the rehearsal of the dream rather than the dream itself that is later remembered. Just why they are forgotten is something of a mystery, since dreaming activates the hippocampus, which is the hub of the memory system. One suggestion is that the prefrontal lobe of the brain, which plays an executive role in memory formation, is deactivated during dreaming. Another is that the brain is in a different chemical state due to deactivation of monoaminergic systems, preventing memory formation. Or perhaps the hippocampus is active precisely because it is involved in consolidation, organising the formation of past memories, but not establishing a memory for the dream itself.

Whatever the case, the lack of memory for dreams themselves must be adaptive, because it would not serve us well to confuse our dreams with what has actually happened—although occasionally we do this anyway. Although mostly forgotten, dreams sometimes create an elusive sense of mystical or cosmic presence that can persist, as captured by the Victorian poet Alfred, Lord Tennyson in his poem 'The Two Voices':

'Moreover, something is or seems,
That touches me with mystic gleams,
Like glimpses of forgotten dreams—

'Of something felt, like something here;
Of something done, I know not where;
Such as no language may declare.'

The cycles of REM sleep are orchestrated by a structure deep in the brainstem called the 'pons' (Latin for 'bridge'). REM dreams are dominated by vision, even though the eyes are closed and vision is normally precluded anyway by darkness. Roughly half also include an auditory component, around 30 per cent feature sensations of movement or touch, and hardly any involve taste or smell. We can dream of walking or running, and even people who have suffered paralysis of the lower half of the body dream of moving freely. Our actual movements, though, are inhibited during sleep. This too was probably adaptive, especially during earlier times, because the body is especially vulnerable to attack while we sleep, and movement might alert night-time predators. The inhibition of movement also prevents us from acting out our dreams in the real world, with potentially disastrous consequences, and it sometimes affects the dream world as well. The sleep researcher Allan Hobson refers to 'the annoying flaccidity of our legs as we try to run faster and faster to elude the imaginary dream assailant'.

REM sleep emerges in the foetus, and peaks in the third trimester, when it's there all the time. But the foetus probably doesn't dream in any meaningful sense of the word. A little later, but still before birth, NREM sleep and wakefulness join the cycle, and after birth the newborn spends roughly equal time awake, in REM sleep, and in NREM sleep. REM sleep gradually drops away, but settles to about 1.5 hours per night throughout most of our lives—roughly the time you might otherwise spend watching a movie or a TV soap. Dreaming itself, though, may be slow to develop. Preschoolers do dream, but reports suggest that their dreams are simple and static, without emotion and with no involvement of the dreamer. Night terrors experienced by some children are probably not the result of bad dreams, but seem to be induced by the disorientation caused by not being properly awake. David Foulkes found that children under the age of seven report dreams when awakened from REM sleep only 20 per cent of the time, compared with 80 to 90 per cent in adults. Dreams might roughly parallel the development of mental time travel itself. As explained earlier, it's not until the age of about four that the child can mentally escape the present and envisage coherent scenarios in which they are somewhere else at some other time. At around the age of seven, dreams develop a narrative quality, and incorporate characters that move around, including the child as the dreaming self.

This slow development of dreaming raises the question of whether non-human animals dream. Many of them do have REM sleep, though. In birds, it seems that only hatchlings go into REM sleep. NREM sleep emerged only in land animals, going back at least to the origin of mammals nearly 200 million years ago, with REM sleep kicking in at around the time marsupials split off some 150 million years ago. But REM sleep really took hold with the emergence of placental mammals from around 50 million years ago. Kangaroos are engaged in REM sleep for only about a tenth of the time that we are. And dog owners claim that their pet dogs dream because of the faint twitching and small noises they sometimes make while asleep by the fire, but we can only guess what they might be dreaming about. It is unlikely, though, that their dreams have the narrative quality of human dreams—although we saw in Chapter 4 that rats seem to dream of perambulating through mazes. More of that later.

REM sleep is not just a dream machine, but seems to be critically important in regulating temperature. Birds and mammals are warm-blooded creatures, and their body temperatures are internally controlled. The system of temperature control, though, seems to depend uniquely on adequate REM sleep. Rats totally deprived of sleep, or even just of REM sleep, all died through a failure of metabolism and thermoregulation. This could mean that REM dreams are simply an epiphenomenon, a secondary consequence of REM sleep, and of no importance in themselves. They are visited upon us for free, as it were, as a car dealer might throw in a sound system when selling you a new car. Even so, people have always sought meaning in the semi-random hustle of images and feelings that occupy our dreams, although this may be no more valid than finding significance in the pattern of tea leaves or the alignments of the planets.

The ancient scholars believed dreams were inspired by gods and demons. They also believed they foretold of the future, an idea that persists, and indeed seems almost irresistible. Abraham Lincoln is said to have dreamed about an assassination two weeks before he was shot dead, and Mark Twain told of a dream in which he saw his brother's corpse lying in a coffin a few weeks before he was killed in an explosion. People often claim to have had premonitions of major tragedies. In 1966, in the small Welsh village of Aberfan, heavy rains caused a landslide which smashed into the village school, killing 139 children and five teachers. John Barker, a British psychiatrist interested in the paranormal, arranged for a newspaper to ask whether any of its readers had had a premonition of the disaster, and received 60 letters from across England and Wales. Over half of them claimed to have had the premonition in a dream.

It is unlikely that these premonitions are evidence of the paranormal. They may be based simply on knowing that the weather was bad, and the memory for dreams can be later embellished to fit events after they occur. As we have seen, memory for dreams is in any case poor and fragmentary, and as much the stuff of fabrication as of true recollection. As I explained in Chapter 2, even the memory of everyday events is more like a story than a tape recording, or a video. Some people, Eeyore-like, seem to expect disasters to happen any moment, and no doubt dream about them too. Sooner or later their dreams are likely to be fulfilled.

Bob Dylan, at least, in his song 'I Feel a Change Comin' On', is skeptical. He reckons dreams never work for him even when they *are* true. You've got better things to do than dream.

Enter Sigmund Freud

Sigmund Freud, father of psychoanalysis, regarded dreams as the product not of gods or demons, but of the unconscious underworld of the mind. The unconscious harbours the disturbing thoughts, say of sex, fear, aggression, or even murder, that arise from our animal instincts, and which society demands that we suppress. The object of psychoanalysis is to reveal the hidden thoughts of the unconscious so that patients can face the true origins of their neuroses. Dreams, wrote Freud, are 'the royal road to the unconscious'; they provide a glimpse of thoughts that are normally censored. Even so, those thoughts are still disguised in symbols, which need to be interpreted to reveal what they hide. In Freud's world, at least, thoughts of forbidden sex seemed to predominate. Weapons or tools may be symbols of the male organ. Going up and down ladders or stairs is symbolic of the sexual act. Complicated machines are 'very probably' the male genitals, as are landscapes, 'especially those that contain bridges or wooded mountains'. 'Tables, whether bare or covered, are women.'

The problem, as many have pointed out, is that one can interpret almost anything in sexual terms, and form conclusions that are based more on whim than on the true nature of things. I have tried in vain to think of any object or activity that could *not* be interpreted sexually. (Suggestions welcome.) Freud's view that suppressed sexual misadventure underlies neuroses was also a forerunner of the therapies that surfaced in the 1980s and 1990s, and mentioned in Chapter 2. It again exposes the logical error of affirming the consequent. An unfortunate experience with women, for instance, may possibly result in dreams about a table, but one may well dream about a table as a result of an experience with, well, a table. I have it on reasonable authority that people sometimes dream about sex itself, without symbolic cover.

That said, Freud's *Interpretation of Dreams*, published in 1900, is an insightful and scholarly account of theories about dreams. And he was himself not entirely confident of his interpretations; in a letter to his friend Wilhelm Fliess in 1906 he wrote:

> Do you suppose that some day a marble tablet will be placed on the house, inscribed with these words: 'In this house on July 24th, 1895, the secret of dreams was revealed to Dr. Sigm. Freud'? At the moment I see little prospect of it.

Freud also mentions what he calls 'typical dreams', which are dreams that recur and seem to be universal. They include dreams of falling from a great height, flying, or being naked. Freud suggests that dreams of falling or flying hark back to childhood, when one is carried and maybe playfully thrown in the air by a parent, or ridden on a playground swing or see-saw. That seems innocent enough. Dreams of nakedness, he says, express a normally suppressed tendency to exhibitionism, but are accompanied by a sense of shame, heightened by the inability to hide one's nakedness 'by means of locomotion'. Does this sound familiar? He goes on to say: 'I believe that the great majority of my readers will at some time have found themselves in this situation in a dream.'

But perhaps dreams of nakedness derive from being naked as an infant, or reflect the fear of being caught with one's pants down. Another typical dream, also mentioned by Freud, is that of failing an exam, or being required to repeat a course. Freud suggests that this relates to anxiety over early misdeeds, but later expressed in terms of more contemporary fears. Even so, I still occasionally dream of failing an exam, or more commonly of not having done any exam preparation, but I haven't had to take an exam for some 50 years. Nevertheless, our early anxieties do seem to persist in dreams. I still dream with trepidation of being back in boarding school, but that's not a fear that haunts me now in my day-to-day existence (and it wasn't really that bad). I also dream of being lost in a strange city, which I suppose could happen one of these days—but I don't lose any sleep over it. Whatever the origins of these typical dreams, though, their universality does suggest they are not merely random, kaleidoscopic jumbles.

The Freudian idea that dreams are symbolic disguises of shameful or forbidden thoughts has largely lost favour. He was probably right, though, in recognising the unconscious mind, which seems to play a role when consciousness is focused elsewhere. We all know of the 'Aha!' experience that pops into our heads well after a conscious attempt to solve a difficult crossword clue or remember someone's name. The mathematician Henri Poincaré once described how he came upon one of his insights while on a geological excursion:

> The incidents of the journey made me forget my mathematical work. Having reached Coutances, we entered an omnibus to go some place or other. At the moment when I put my foot on the step, the idea came to me, without anything in my former thoughts seeming to have paved the way for it, that the transformations I had used to define the Fuchsian functions were identical with those of non-Euclidian geometry!

Tyger! Tyger! Burning bright
In the forests of the night,
What immortal hand or eye
Could frame thy fearful symmetry?
    —William Blake, from *Songs of Experience*

The Finnish psychologist Antti Revonsuo suggests that dreams are simulations of threatening events, providing the opportunity to develop ways of recognising and coping with real-life dangers. Such dreams emerged during the Pleistocene as an adaptive response to an environment fraught with danger. Blake's 'Tyger', then, is a threat from prehistoric life, perhaps not so much from the forests of the night as from the open expanse of the African savannah. The 'typical dreams' referred to above are indeed often threatening, sometimes nightmarish. Dreams do seem to have something of a primeval character—we don't seem to dream of reading, writing, using a computer, even driving a car. Revonsuo suggests that the dream system harks back to times no longer relevant in the modern world, but is nonetheless ingrained in emotional memory. Dreams seem to have much in common with children's stories, which are alive with animals, forests and dangerous things. I am led to wonder, in fact, whether we have recreated a primeval world for our children, providing them with the stuff of bad dreams for the rest of their lives.

Revonsuo's theory has prompted analyses of a large number of dreams, collected in several different countries. Some two-thirds to three-quarters of dreams include threatening events, which is much higher than the proportion of threats contained in parallel logs of events during waking hours. The threats experienced in dreams are also much more severe. Nevertheless, people actually exposed to threats or real-life traumas have more dream threats than those who lead more tranquil lives. One study comparing dreams across different countries showed the proportion of dream threats to be lowest in Finnish children, at just under 40 per cent. According to the authors of this study, these children had lived all their lives in the most peaceful and stable environment of all the studied children—and perhaps weren't told scary stories. Among traumatised Kurdish children, in contrast, the proportion was 80 per cent.

The most common forms of dream threats, at around 40 per cent, had to do with aggression, while the rest were made up largely of failures, and of accidents and misfortunes. Echoing my own dreaming of the fear of exams, the threats encountered in dreams derived more from old memories than from recent events. It appears that the emotional significance of the threat was more potent than its recency. Most dream threats were directed at the dreamer, but in about 30 per cent of cases the threat was directed at significant others, such as close kin, friends and allies.

The idea that threatening dreams hark back to the Pleistocene does carry some plausibility. In the words of Thomas Hobbes' *Leviathan* (1651), early life was 'nasty, brutish, and short', and evidence from fossil remains from the Pleistocene indeed reveals a dearth of individuals over the age of 40, in Neanderthals as well as in early humans. For our hunter-gatherer forebears, threats to life must have included dangerous predators and perilous means of getting food. It may well have been adaptive to dream of simulated attacks, and so develop coping strategies. This is not to say that memories of tigers and other predators are encoded in the genome itself, although the sense of danger in unfamiliar places or at the sight of unfamiliar creatures may well have become ingrained in our biological make-up from more threatening times. In children's stories and cartoons, we seem to do our best to revive the Pleistocene. But not all threats are present in dreams. Another potent threat to life in the Pleistocene was disease, but dream life is not well equipped to find cures for illness or infection. We seldom dream of being ill, and even if we did, there isn't much the dream could do about it. Dreaming seems calibrated to expose threats in which the dream itself can lead to potential solutions.

Like Freudian theory, the threat theory seems to suggest that REM dreams, at least, are peculiar to humans or to our Pleistocene forebears, although perhaps tigers had gleeful dreams of threatening rather than being threatened—I should add that the tables have been turned, and the tiger is now the threatened species. The threat theory may be more general, though. REM dreams are driven by processes in the brainstem, which probably well up through the emotional centres before influencing higher areas that carry memories. Our human emotions may have characteristics born of the dangerous environment of the Pleistocene, but emotions themselves have much more ancient origins. Walter Ratty, too, may dream of dangerous cats.

In his 1872 book *The Expression of the Emotions in Man and Animals*, Charles Darwin suggests that there is only one emotional expression that seems to be unique to humans. 'Of all expressions,' he writes, 'blushing seems to be the most strictly human'. Now that's something I'd never have dreamed of.

The return of Walter Ratty

The threat theory is based largely on dreams occurring during REM sleep, which are the most vivid and memorable, and also the most frequent. Dreams during NREM sleep, and especially early-night NREM sleep, seem to tell a different story, in which dreams represent recent experiences rather than old emotional fears. In this respect, NREM dreams seem more in line with hippocampal recordings, whether from rats or humans. In Chapter 4, I described how 'ripples' of neural activity in the rat hippocampus correspond to trajectories in a familiar terrain, such as a maze. The trajectories may correspond not only to paths actually taken, but also to new paths, perhaps in anticipation of further exploration. These ripples occur both while the animal is awake as well as when it is asleep. It is during early NREM sleep that reactivation of trajectories is strongest.

Erin Wamsley and Robert Stickgold have studied dreaming in humans during NREM sleep, and around half of these dreams include at least one reference to a recent experience during waking hours. In only 2 per cent of cases, though, did the dream replay the experience as it actually happened. Here's an example of how a dream can reflect aspects of the experience without actually duplicating it:

Waking memory source: When I left Starbucks [at the end of my shift], we had so many leftover pastries and muffins to throw away or take home. I couldn't decide which muffins to take and which to toss...

Corresponding dream report: My dad and I leave to go shopping. We go from room to room, store to store. One of the stores is filled with muffins, muffins, muffins from floor to ceiling, all different kinds, I can't decide which one I want...

REM dreams can be quite prolonged, and probably occur in real time, unfolding as they would if they were happening in the actual world, whereas NREM dreams flit by in a fraction of a second, at least as judged from the ripples in the rat hippocampus. It is well established that NREM sleep is important for the consolidation of learning. In one study, people were trained to navigate a virtual maze, and did much better following an afternoon nap in which they dreamed about the maze than if they dreamed of something else. Their after-nap performance was unaffected by waking thoughts about the maze. Best get a good night's sleep after studying for that exam.

In this study, too, the dreams were not exact replays. Two participants reported hearing the music associated with the maze task, but did not dream of the maze itself; three others described other maze-like environments. Part of the consolidation process, then, seems to go beyond an exact experience so that people— and rats—gain a wider understanding of what they have learned, and perhaps greater adaptability to the future. I suggested in Chapter 4, though, that these dreams, and their daytime equivalents, are also the basis of mental time travel, with an eye to the future rather than the past.

Our brains, then, are far from dormant during the night. With the absence of sensory stimulation and paralysis of movement, and under the cover of darkness, they seem to take the opportunity to service the mind as well as the body, just as your car needs to be serviced from time to time to check the wearable parts and tune its performance. Mental servicing includes emotional regulation during REM sleep, and the consolidation and expansion of our memories during NREM sleep. The two are also distinguished in terms of the chemistry of the brain. One such difference has to do with acetylcholine, which is a neurotransmitter—that is, it influences the efficiency with which neurons in the brain connect with one another. Acetylcholine is at a minimum during NREM sleep, and is also reduced in 'quiet wakefulness', when people, and perhaps rats, are apt to daydream. Lowering of acetylcholine is thought to promote consolidation of memories by supporting the flow of information from the hippocampus to other parts of the brain, where the details of memory are stored. In REM sleep, in contrast, the levels of acetylcholine are higher than they are when we're awake, which is perhaps another reason why we find it virtually impossible to remember REM dreams.

Much of the theory of dreams seems to focus on the negative— the threats, the traumas, the failed exam, the harping on past misfortunes and embarrassments. We should remember, though, that many of our dreams are positive, and fun to be part of. Dr Seuss, whose children's stories are themselves a bit dream-like, was quoted as saying: 'You know you're in love when you can't fall asleep because reality is finally better than your dreams.' This morning, my four-year-old granddaughter tells me she dreamt of Pooh and Tigger. But she enjoyed telling it—the Tigger of A. A. Milne's *House at Pooh Corner* doesn't have the fearful symmetry of William Blake's Tyger. And as bears go, Pooh is as lovable and cuddly as anyone's teddy bear.

Dreams may seem a wasted form of mind-wandering, since we forget nearly all of them. Rather than providing conscious nutrients for our waking lives, they may serve rather to activate the unconscious, to create the internal terrains for later mental meanderings. Occasionally, we do remember our dreams, and sometimes these form the basis for creative ideas, as I shall illustrate in the final chapter. First, though, I consider another form of mind-wandering that, although largely beyond our control, can remain in memory and affect our conscious lives.

8.
HALLUCI–
NATIONS

• • •

> A hallucination is a fact, not an error. What is erroneous is a
> judgement based upon it.
>   —Bertrand Russell, from *Theory of Knowledge*

The word 'hallucination' was first used in the early sixteenth
century simply to mean 'a wandering mind', and what we
now understand as hallucinations were called 'apparitions'.
In Shakespeare's *Macbeth*, for example, the three witches summon
up three apparitions, so named, to warn Macbeth that Macduff is
coming back to Scotland to ruin him. The third apparition is a child
holding a tree and wearing a crown, and declares:

> Macbeth shall never vanquish'd be until
> Great Birnam wood to high Dunsinane hill
> Shall come against him.

Apparitions, like dreams, were thought to tell of the future, and
later in the play Birnam wood was indeed seen as though moving to
Dunsinane hill, contributing to Macbeth's downfall.

The term 'hallucination' was introduced with the meaning we
use today in the 1830s by the French psychiatrist Jean-Étienne
Esquirol, and from that point linked with psychiatric disorder,
at least until recently. Hallucinations are still examples of mind-
wandering, though, since they are produced in the mind, and
are imposed as 'extras' on present reality. They are most simply
defined as perceptions of things that are not there, most commonly
as visions or voices, but occasionally through other senses, such

as smell or touch. Macbeth's third apparition was both seen and heard, unlike children of the Victorian era who were supposed to be seen but not heard.

Hallucinations are perceived as though they are real, but differ from perceptions of reality in that they are not perceived by anyone else. They are not like normal memories of past events or your everyday mind-wanderings since they take place in the here and now. Where our normal mental time travels tend to be indistinct and contained within the mind, hallucinations are projected onto external space, with the vividness of reality. Sometimes, hallucinations do interact with real-world perception, superimposing unreal events onto an otherwise normal world. In his 2012 book *Hallucinations*, Oliver Sacks gives the example of looking at someone in front of you and seeing not just one person, but five identical people in a row.

Hallucinations can seem so realistic that they are often incorporated into normal routine. Sacks tells of a hallucination he experienced while under the influence of hallucinatory drugs. He heard his friends Jim and Kathy knock on his door, and called out to welcome them into the living room, and chatted with them while he cooked them ham and eggs. When the meal was ready he put it on a tray and walked into the living room. There was no one there. The conversation had seemed entirely normal, with Jim and Kathy speaking in their normal voices, but the episode was a hallucination.

Although the great majority of hallucinations are seen or heard, they sometimes inhabit other senses. William James writes of a close friend who had a vivid experience of being touched on the arm so realistically that he searched the room for an intruder.

More elusive is what James described as a 'sense of a presence', the feeling that there is another person in the room. Although not impinging on the actual senses, the ghostly intruder may be felt to be facing in a particular direction, and even located in a particular spot. Hallucinations of this kind may well be interpreted as the presence of God.

Hallucinations have often been taken to indicate mental disorder, with auditory hallucinations at one time suggesting referral to the psychiatrist and visual hallucinations to the neurologist. In 1973, eight people without psychiatric symptoms conducted an interesting experiment. They presented themselves at various hospitals in the United States, and complained, falsely, that they 'heard voices'. Although they acted normally in other respects, they were diagnosed with mental illness, seven with schizophrenia and one with manic-depressive psychosis, and were admitted to psychiatric wards. Needless to say, they were eventually released to tell the tale. This is another example of the logical error of affirming the consequent. People with psychiatric disorders may indeed be prone to hallucinations, but hearing voices that are not there need not imply psychosis. Indeed, the majority of people who hear voices are not psychotic. Some people go out of their way to experience hallucinations, perhaps simply for pleasure, or in the belief that they enhance creativity. The Bohemian writer René Karl Wilhelm Johann Josef Maria Rilke (better known as Rainer Maria Rilke) waited years for the Voice to speak so he could transcribe what it said, a delay perhaps caused by the Voice struggling to address Rilke's surfeit of names.

Until Esquirol's pronouncements in the eighteenth century, hearing voices was regarded as quite normal, and attributed to gods or demons. Even in more recent times, voices are sometimes heard as messages or commands from God, occasionally leading to religious conversion. One case, quoted by William James in his *Varieties of Religious Experience*, was George Fox, founder of the Quaker religion. He happened to be walking with friends near the cathedral city of Lichfield in England, when he was called:

> Immediately the word of the Lord came to me, that I must go thither. Being come to the house we were going to, I wished the friends to walk into the house, saying nothing to them of whither I was to go. As soon as they were gone I stept away, and went by my eye over hedge and ditch till I came within a mile of Lichfield where, in a great field, shepherds were keeping their sheep. Then was I commanded by the Lord to pull off my shoes. I stood still, for it was winter: but the word of the Lord was like a fire in me.

Such voices, it has been suggested, may be a lingering relic of an ancient era when people paid much more attention to the voices of the gods than most of us do today.

Bicameral minds

Julian Jaynes, in his best-selling book *The Origin of Consciousness in the Breakdown of the Bicameral Mind*, published in 1976 but still in print, argued that up until around 1000 BC, or earlier, people were effectively controlled by hallucinations, interpreted as the gods issuing commands. Jaynes finds evidence for this in the ancient Greek epic the *Iliad*, set during the Trojan War in the twelfth century BC. It was maintained in oral traditions until eventually written down, supposedly by Homer, around 850 BC. People are described in the *Iliad* as though having no sense of conscious self; there is no first-person narrative. As Jaynes puts it: 'the gods take the place of consciousness'. He goes on to write of the 'bicameral mind', divided between gods who give instructions and the person who hears them. '[I]t is highly probable', he writes, 'that the bicameral voices of antiquity were in quality very like such auditory hallucinations in contemporary people. They are heard by many completely normal people to varying degrees.'

The bicameral mind began to break down as a result of catastrophic events in the second millennium BC. Wars, floods and migrations meant that 'chaos darkened the holy brightnesses of the unconscious world'. The transition from passive reliance on the voices of the gods gradually gave way to individual responsibility for action. According to Jaynes, the shift is strikingly evident in the *Odyssey*, sequel to the *Iliad*, and also attributed to Homer, but very different in style and in depiction of the mind. In the *Odyssey*, the dominance of the gods recedes, and the characters become capable of making their own decisions. They speak in the first person.

Jaynes also suggests that the shift is accompanied by a switch in brain dominance. In the bicameral mind, it was the right side of the brain that received the voices of the gods and relayed them to the left side, which is specialised for language and so 'hears' and obeys what the gods are saying. With the breakdown of the bicameral mind, dominance flipped to the left side, which now asserts control over action. Even so, hallucinations still persist, with reduced frequency, as vestigial functioning of the right brain.

Jaynes died in 1997 but remains something of a cult figure. Whether his theory makes historical or neurological sense is doubtful. The time gap between the *Iliad* and the *Odyssey*, whether in terms of the events they depict or the times at which they were written, seems too narrow for such epochal changes. Jaynes' account is based on events that were geographically constrained even by ancient standards—what of the populations of Asia, the Americas, Australia? Differences between the two sides of the brain have populist appeal, but are often more to do with folklore than neurological fact. The right and left brains have become little more than convenient pegs on which to hang our polarities—emotion and reason, love and war, female and male, and in an ironic reversal the left and right of politics. And yet the left brain/right brain story continues, most recently in Iain McGilchrist's remarkable book *The Master and his Emissary*, published in 2009, in which the right brain is the master and the left brain the mere emissary. McGilchrist argues that control should be transferred back from the left brain to the right—a return, perhaps, to the bicameral mind.

But there could be some truth to the claim that hallucinations come mainly from the right brain.

Electrical induction

One way to induce hallucinatory or dream-like experiences is to open up the skull and apply weak electrical stimulation to the exposed brain. This is not recommended as a party game, but can prove useful as a preliminary to brain surgery. Wilder Penfield was a neurosurgeon who established the famed Montreal Neurological

Institute, and pioneered the use of surgery to remove brain areas containing the sources of epileptic seizures. Prior to surgery, he often applied mild electrical stimulation to the exposed brain, so that he could gain a sense of which portions could be surgically removed. Patients are conscious and able to talk during brain operations, and if stimulating a particular region prevented a patient from talking, then that region was assumed critical to speech, and should not be excised. To Penfield's surprise, though, the stimulation sometimes caused patients to report hallucinatory or dream-like experiences.

These 'experiential responses', as Penfield called them, were always induced by stimulating the temporal lobe, never by stimulating the other lobes. The temporal lobes, as those very lobes may remind you, house the hippocampus on the inner surface, and are therefore critically involved in memory. The experiential responses often seemed to Penfield, and to the stimulated patients, like replays of earlier memories, leading Penfield to suggest that we store away in our brains much more than we can voluntarily recall—perhaps even that *everything* we experience is tucked away in our memory banks. This interpretation helped fuel the idea in psychotherapy that harmful memories are repressed, and need to be coaxed out of the brain for therapy. As I suggested in Chapter 2, this notion can be dangerous, sometimes leading to false memories being unwittingly implanted by overzealous therapists.

The alternative interpretation is that the experiential responses were more like dreams, or hallucinations, than replays of the past. One woman patient claimed she saw herself in childbirth, and felt as though she were reliving it. A twelve-year-old boy said he saw robbers coming at him with guns, a vision which Penfield suggested came from his reading of comics. A 45-year-old woman saw the faces of two former schoolteachers, who moved toward her and crowded her, and she cried out in terror. A fourteen-year-old girl saw herself as a seven-year-old walking through a field of grass and sensing that a man was behind her, about to smother her or hit her on the head. This last experience resembled a dream that she had had at age seven in which she was walking through a field with her brothers when a man approached from behind and asked how she would like to get into his bag with the snakes and be carried away. These experiences seem more like simulated threats than replays of actual events.

In many cases the patients heard the voices of people they knew, or familiar tunes, but there was little indication that these were exact replays. When asked whether hallucinated episodes had actually occurred at some earlier period in their lives, some patients replied that they thought they might have, but there was little suggestion of certainty. Some hallucinations were nevertheless vivid and detailed, as in the following from a 26-year-old woman:

> She had the same flash-back several times. These had to do with her cousin's house or the trip there—a trip she has not made for ten to fifteen years but used to make often as a child. She is in a motor car which had stopped before a railway crossing. The details are vivid. She can see the swinging light at the crossing. The train is going by—it is pulled by a locomotive passing from the left to right and she sees coal smoke coming out of the engine and flowing back over the train. On her right there is a big chemical plant and she remembers smelling the odour of the chemical plant.

The windows of the automobile seem to be down and she seems to be sitting on the right side and in the back. She sees the chemical plant as a big building with a halffence next to the road. There is a large flat parking space. The plant is a big rambling building—no definite shape to it. There are many windows.

Even in this case, though, she says she doesn't know whether this was a replay of an actual event, but she sees it as though it was. What seems more likely is that it was a montage built from elements of memory.

All of these patients had histories of epilepsy, and some of the experiences also occurred during seizures, while others appeared only during brain stimulation. Roughly equal numbers of patients were operated for left- and right-temporal lobe epilepsy, but only 40 out of 520 had hallucinations induced electrically. In 25 of the 40, the epilepsy was focused in the non-dominant side of the brain—in most cases the right side. This does give some support to Jaynes' idea that it's the right brain that harbours hallucinations, although it was mainly the visual hallucinations that favoured the right brain, with auditory hallucinations as often from left as from the right. When speaking, at least, it seems that Jaynes' gods may have visited through either side.

These hallucinatory experiences no doubt draw on memory, and activation of the hippocampus surely played a major role. But this is not to say they replayed memories as they actually occurred. More often, they seem to be formed from elements of remembered places, people, or things, but constructed in ways that need bear little relation to earlier happenings, and indeed are often manifest in bizarre ways that could not possibly have actually happened. They are more like dreams than the mental time travels of our normal lives, and even those are in part fabricated. We probably never remember things precisely as they actually happened.

Sensory deprivation

Hallucinations frequently occur when normal sensory input is removed or reduced, as though the brain invents an imaginary world when the real world is shut off, just as it does when you dream at night. One form of sensory deprivation is blindness, and people who have lost their sight often have visual hallucinations. These make up what is known as Charles Bonnet syndrome, named for the Swiss naturalist who became interested in the 'visions' his grandfather Charles Lullin experienced as his eyesight deteriorated. These hallucinations were quite florid. Once, when two of his granddaughters came to visit, Lullin saw two young men appear in magnificent red and grey cloaks, and wearing hats trimmed with silver. When he exclaimed on their presence, his granddaughters said they saw nothing, and the two men dissolved. As my eyesight deteriorates, I expect to see my twin granddaughters, now aged four, with equally handsome young men in attendance—but they'd better be real.

Charles Bonnet syndrome was once considered rare, but it is now known that some 15 per cent of elderly patients with deteriorating vision have complex hallucinations, made up of people, animals, or scenes. As many as 80 per cent see more diffuse shapes, colours, or patterns, which perhaps arise from random activity in the visual cortex itself. Deprived of normal input, the restless visual brain stirs up its own mischief.

Deafness can also result in hallucinations, typically of music, but occasionally of other sounds, such as birdsong, bells chiming, a lawnmower. Unlike visual hallucinations, musical hallucinations are generally true to reality. They can be highly detailed, with every note and every instrument distinctly heard, although sometimes only a few bars are hallucinated, played over and over. Oliver Sacks mentions a patient who heard part of 'O Come, All Ye Faithful' nineteen and a half times in ten minutes, timed by her husband. Another patient, a violinist, hallucinated one piece of music while actually performing another piece at a concert. These hallucinations, though, are not so much replays of previous episodes as repeats of well-known and no doubt often repeated experiences.

Hallucinated music is like the earworms mentioned in Chapter 1, persistent and hard to dispel, but generally more vivid and true to life, with a level of detail and accuracy that may astonish the hallucinator, who in some cases is normally unable even to hold a simple tune. The seeming reality of hallucinated music is illustrated by a woman who wrote to Sacks:

> I kept hearing Bing Crosby, friends and orchestra singing 'White Christmas' over and over. I thought it was coming from a radio in another room until I eliminated all possibilities of outside input. It went on for days, and I quickly discovered that I could not turn it off or vary the volume.

Another partially deaf 60-year-old woman heard persistent music as though from a radio at the back of her head, including one song that played repeatedly for three weeks before another song took over. She did not even recognise many of the songs she heard, but was able to hum the tunes, which were then identified by members of her family. These songs were evidently buried deep in her memory, but were somehow able to surface only as hallucinations. Aside from her partial deafness, she showed no evidence of neurological or physical disorder.

You don't need to be blind or deaf to suffer from sensory deprivation. People held in cells or dungeons may seek some consolation in what Sacks calls 'the prisoner's cinema', made up of hallucinations and dreams. Visual monotony can also do the trick. Sailors, polar explorers, truckers and pilots are all prey to visual hallucinations, which can sometimes be hazardous. The poet Samuel Taylor Coleridge, himself somewhat driven by drug-induced hallucinations, captures something of the hallucinatory life of a sailor in his 1798 poem *The Rime of the Ancient Mariner*: 'Yea, slimy things did crawl with legs / upon the slimy sea.'

At McGill University in the 1950s, researchers paid people to stay in soundproofed cubicles, wearing gloves and translucent goggles to cut down stimulation, for as long as they could bear it. At first they fell asleep, but when they awoke became increasingly bored, craving stimulation. Soon their brains provided it, with hallucinations that grew progressively more complex, culminating in quite elaborate scenes. One saw a procession of squirrels marching across a snowfield; another prehistoric animals walking about in the jungle. In later studies, volunteers were floated in tanks of warm water, effectively removing all sensory input. This austere environment rapidly induced hallucinations, and in the 1970s the tanks became avidly sought after as consciousness-expanding drugs.

Hallucinations induced under sensory deprivation can also be shown to differ from normal visual memories in terms of the brain areas involved. A group of researchers in Germany persuaded a woman artist to be blindfolded for 22 days, so that she experienced visual hallucinations. While blindfolded, she also had several sessions in an MRI scanner, and was able to indicate when her hallucinations came and went. The scans revealed activity in her visual system precisely linked to the hallucinations. Afterwards, she drew illustrations of some of the hallucinations, and when she was asked to conjure them up in imagination the visual areas were not activated. In the absence of visual input, we seem to be unable to intentionally activate the parts of the brain that give us truly visual experiences, but hallucinations can do it for us.

Drugs

The fastest way to hallucinate is to take hallucinatory drugs—'transcendence on demand', as Oliver Sacks puts it. So great is the human flirtation with drugs that we have evolved a symbiotic relationship with nearly 100 plants with psychoactive substances. It seems they may need us as much as we need them, perhaps not so much for their mind-expanding properties as for the sense of euphoria they supply. We should not take full credit for the existence of these plants, though. Some plants have evolved psychoactive agents to deter predators, or to entice other animals to eat the fruit and spread the seed. And we humans have gone beyond plants to synthesise new hallucinogens.

In the 1890s, western people discovered peyote, also known as mescal, a cactus with psychoactive properties that had probably been used for over 5000 years by Native Americans in religious rituals or as a medicine. One who described its effects was Silas Weir Mitchell, a distinguished American physician. At one point he took a solid dose, went on several house calls, and then settled in a dark room, closed his eyes, and experienced 'an enchanted two hours'. These included vivid arrays of colour and light, a grey stone that grew to a great height and became an elaborate Gothic cathedral, clusters of huge precious stones or coloured fruits—'all the colours I have ever beheld are dull as compared to these'.

In his 1902 book *The Varieties of Religious Experience*, William James refers to the case of a Mr Peek, perhaps a forebear of the savant Kim Peek, who wrote as follows of his mescal-induced experience:

> When I went in the morning into the fields to work, the glory of God appeared in all his visible creation. I well remember we reaped oats, and how every straw and head of the oats seemed, as it were, arrayed in a kind of rainbow glory, or to glow, if I may so express it, in the glory of God.

Mescal and other hallucinogens seem to zero in on regions of the brain involved in vision, and especially the perception of colour, as well as inviting religious experience.

Oliver Sacks was himself an enthusiastic player in the drug culture of the 1960s, the era of the Beatles' song 'Lucy in the Sky with Diamonds', composed as a celebration of the drug lysergic acid diethylamide (LSD). He started with cannabis, which gave him an experience that was 'a mix of the neurological and the divine'.

He moved on to Artane, a synthetic drug allied to belladonna, and it was this that produced the hallucinatory visit of his friends Jim and Kathy, described earlier. Shortly after he had eaten the ham and eggs he had cooked for his absent guests, he heard a helicopter, bringing his parents for a surprise visit. Amid the deafening roar of the helicopter landing beside his house, he quickly took a shower and changed his clothes. You know the rest—there was no helicopter, no parents. Just a sad Sacks.

He later developed a cocktail of drugs made up of amphetamine, LSD and a pinch of cannabis. He wanted especially to see the colour indigo, which Isaac Newton had rather arbitrarily included in the colour spectrum. After taking his cocktail he faced a white wall and demanded: 'I want to see indigo—*now!*' He was rewarded by 'a huge, trembling, pear-shaped blob of the purest indigo'. It was, he thought, the colour of heaven. This seems to have been a rare example of a hallucination at least partly under the control of the hallucinator—he asked for it, and got it. But his continued experimentation with drugs turned the heaven into hell. The hallucinations themselves turned unpleasant and frightening, his sleep was disturbed, and he developed delirium tremens. With the help of his friend, the American actress Carol Burnett, he managed eventually to escape the addiction and pursue his career as a successful author and neurologist.

Hallucinations are indeed remarkable in expanding consciousness. So are dreams, which might themselves be regarded as hallucinations wrought by sensory deprivation, although their connection with the rhythms of nocturnal eye movements suggests that they are natural events. All cultures have had their obsessions with hallucinogenic agents, as though there is a human imperative to explore the boundaries of the mind beyond what life normally provides. It is hallucinations, perhaps, that fuel religions, providing a sense that there is more to existence than the daily routine, and the foreboding that life is temporary. '[A]ll our yesterdays', says Macbeth in Shakespeare's play, 'have lighted fools / The way to dusty death.'

Drug-induced hallucinations seem largely visual, but an exception is provided by Evelyn Waugh in his semi-autobiographical novel *The Ordeal of Gilbert Pinfold*. Waugh was a heavy drinker, and in the novel his alter-ego Gilbert Pinfold tries to cure his woes by adding a strong sleeping potion of chloral hydrate and bromide to his regular intake of alcohol. He then decides to take a restorative cruise to India. He runs out of the sleeping potion but continues to drink heavily. Then the hallucinations begin. They are exclusively auditory—mostly accusatory voices, but also music, a barking dog, a murderous beating, and phantom shipboard sounds. The hallucinations become increasingly preposterous, and are accompanied by delusions that his persecutors have machines than can read and broadcast his thoughts. All the while, though, he sees the world and his shipboard surrounds as normal.

In western society at least, alcohol has been for the most part the drug of choice. William James commented that 'the sway of alcohol over mankind is unquestionably due to its power to stimulate the mystical faculties of human nature, usually crushed to earth by the cold facts and dry criticisms of the sober hour'. James was evidently referring to the effects of drinking to excess; a glass of good red wine seems harmless enough, and perhaps even good for your health. Prolonged heavy drinking can take you beyond mysticism to delirium tremens, with uncontrolled trembling and hallucinations, as the tale of Gilbert Pinfold suggests. And even when you stop drinking, the withdrawal effects can persist and even grow more extreme. Cheers.

Hallucinations and dreams can take us into regions of the brain that are inaccessible to the conscious will. Our normal mental time travels cannot reproduce the actual experiences on which they are based. By the same token, though, hallucinations don't recapture our exact memories or plans for the future. Perhaps, then, there is a trade-off between the two. During the waking, working hours, we need imagination to be kept in check, so that we don't stray too far from real-world constraints and perish in dreams of immortality or delusions of being superhuman. During the night, or when the sensory world is cut off, the brain takes the opportunity to recharge, and challenge its own limits—much as marathon runners or mountaineers try to challenge their bodily limits. Dreams are a natural part of life's cycle; drugs more of an aberration, an open invitation into a promised heaven but an eventual hell.

It is also remarkable that hallucinations can activate the perceptual systems themselves, so that we see and hear hallucinations as being real. Yet we are accustomed to thinking of perception as driven by the world, through its impact on the bodily receptors—our eyes, ears, noses, organs of touch. The extent to which hallucinations and dreams can invade perceptual systems may suggest that perception is fundamentally driven from within, with information from the world serving merely to guide what we see, hear and smell. Perhaps that's an exaggeration, but hallucinations tell us that there's more to perception than meets the eye.

9.
THE CREATIVITY
OF THE WANDER
ING MIND

• • •

> Creativity is just connecting things. When you ask creative people
> how they did something, they feel a little guilty because they didn't
> really do it, they just saw something. It seemed obvious to them after
> a while. That's because they were able to connect experiences they've
> had and synthesize new things.
>
> —Steve Jobs

The brain is never inactive, the mind never still. For at least half of
our lives, our minds are wandering away from the chores of life—
the homework, the tax return, the board meeting, the meal to be
cooked, even driving the car. During our waking hours, episodes
of mind-wandering arise spontaneously, but we do exercise some
control over where our wanderings take us, whether brooding
over some past incident, planning some future one, working on the
crossword clue, or wondering what our kids are doing or thinking
about. While we're asleep, though, the mind wanders with pre-
dictable frequency in the form of dreams, but with unpredictable
content. We have virtually no control over what we dream about,
although we may mentally act within a dream with some degree of
influence—although as often as not our dreamed intentions seem
to be thwarted. Hallucinations are also dream-like, and we can have
some influence over their appearance, if not what happens in them,
by taking psychoactive drugs or seeking sensory isolation.

Much of our mind-wandering is told in stories. Indeed, if there
is anything special about the human mind it is the capacity to construct intricate narratives, and through language share them with
others. These narratives might be recollections of past events,

plans for the future, or simply made-up stories of often imaginary people in imaginary places doing imaginary things. So it is that the great oral legends that have shaped pre-industrial cultures were born, as well as the epic poems of Homer, the Bible, the plays of Shakespeare, the novels of Jane Austen or Honoré de Balzac, the detective stories of modern times, the soap operas that extend forever on our TV screens. The mind-wandering itself is in the hands or voice of the teller of the tale; the audience or readership is effectively taken on a guided tour. But that guided tour is itself a wandering of the mind into places and times removed from the present.

Mind-wandering has something of a bad press. The wandering mind is said to be an unhappy mind, perhaps even setting us on a path to early death. This view is encouraged by the popularity of mindfulness, and other meditative techniques, designed to focus our thoughts so intently that the mind is tethered into near immobility. The distinction between mind-wandering and mindfulness, though, is not absolute. One of the techniques of mindfulness or meditation is to focus attention on the body, starting with the feet and moving slowly upwards. This is indeed a constrained wander, although hardly a walk in the garden or a stroll along the beach. It may well be that mindfulness is a means of resting the wandering mind, energising its resources. But wander it surely will.

Nature has equipped us to be more than mere robots, locked into specific routines. Our brains have been given the extra resources to escape the here and now and the tasks at hand, and play. Play has evolved because it is adaptive, helping us prepare for life in a complex world. But play itself adds to the complexity, creating a feedback system that enhances our need for further creative play.

Perhaps it is this loop that gave us our extraordinary propensity for mind-wandering and the telling of stories. To continue to survive in the complex worlds we have created, we need to allow the mind to wander—to play, to invent, to be creative.

Creativity

Something I owe to the soil that grew—
More to the life that fed—
But most to Allah Who gave me two
Separate sides to my head.

I would go without shirts or shoes,
Friends, tobacco or bread
Sooner than for an instant lose
Either side of my head.
　　—Rudyard Kipling, from 'The Two-sided Man'

So what of creativity? Let's first dispel the illusion that creativity comes from just one side of our brains—the right side. Google 'right brain' and you get some 660 million hits. 'Left brain', supposedly the dominant half, collects less than half as many, around 270 million. McGilchrist's plea that governance be restored to the right brain, the Master, may be well under way. Googling 'right brain creativity', moreover, gives around 14.5 million hits. The right brain has even wriggled its way into dictionaries. For instance, the fourth edition of the *American Heritage Dictionary of the English Language* (2000) gives the following definitions:

Right-brained *adj*: 1. Having the right brain dominant. 2. Of or relating to the thought processes involved in creativity and imagination, generally associated with the right brain. 3. Of or relating to a person whose behavior is dominated by emotion, creativity, intuition, nonverbal communication, and global reasoning rather than logic and analysis.

Remember, too, that according to Julian Jaynes the gods spoke to us through the right brain.

Kipling's verse, published in 1901, reflects a fascination with the two sides of the brain that followed the discoveries in the 1860s and 1870s that the left brain was dominant for both producing and understanding speech. This led to speculation as to what the right brain might do, and some began to see the differences between the two sides of the brain as complementary, rather than the right being subservient to the left. The left brain came to be seen as the repository of humanity and civilisation, the right as carrying the primitive, animalistic side of our natures. Much interest centred on dual personalities, captured in Robert Louis Stevenson's *Strange Case of Dr Jekyll and Mr Hyde*, published in 1886; Dr Jekyll epitomised the educated, civilised left brain, Mr Hyde the crude, passionate right brain. The two needed to be balanced, because imbalance could cause madness, especially if the right side gained the upper hand, as it were. Associated with madness, and by implication the right brain, was creativity.

This early obsession with the two sides of the brain was largely forgotten from around 1920, but revived in the 1960s and 1970s,[*]

---

[*] We may confidently expect another revival in the 2060s and 2070s.

following work by Roger Sperry and collaborators in California on the split brain. A number of patients suffering from intractable epilepsy had their brains surgically separated down the middle by cutting the corpus callosum, the main tract of fibres that connect the cerebral cortices. With respect to higher-order mental functions such as language, memory, perception, even imagination, then, the two sides of the brain were effectively disconnected, as though two separate minds huddled in the same cranium. The aim of the surgery was to prevent the spread of epileptic discharge from one side of the brain to the other, but it turned out that the outcome was more successful than expected, in many cases leaving the patient largely free of seizures, or with seizures much more easily controlled. Nevertheless, the split brain raised philosophical and psychological questions. Would splitting the brain split the mind? How might the two half-brains differ in their mental faculties?

Roger Sperry and Michael Gazzaniga took the opportunity to find answers. They devised ways to study the mental capabilities of each half-brain independently of the other. Sperry belatedly won the 1981 Nobel Prize in Physiology or Medicine—incidentally echoing the Nobel Prize in Literature won by Kipling in 1907. Sperry and Gazzaniga documented the left-brain specialisation for language, although this had been largely known since the 1860s. Their work also demonstrated spatial and emotional processes for which the right brain seemed dominant. This revived the idea that the two sides were complementary, with the left representing logic and rationality, and the right intuition, emotion and creativity.

As I suggested in the previous chapter, the duality of the brain has been exaggerated, and too often serves to accommodate the polarities that frame our lives. These polarities were driven to some extent by the divisions that fractured social and political life in the 1960s. The left brain stood for the military-industrial establishment of the dominant West, the right for the supposedly peace-loving nations of the East. The women's liberation movement of the 1960s and 1970s also claimed rights to the right brain, the seat of protest against male subjugation—a duality that also goes back to the late nineteenth century when the left brain was also seen as the male side and the right the female side. The dual brain was catapulted into public awareness with the publication in 1972 of Robert Ornstein's best-selling book *The Psychology of Consciousness*.

Emerging from this duality was the idea that the right brain is somehow the engine for creativity—an idea that itself was partly responsible for Julian Jaynes' notion that the gods spoke through the right brain, or Iain McGilchrist's view that the right brain is the master and the left brain the emissary. In 1979, an art teacher called Betty Edwards wrote *Drawing on the Right Side of the Brain*, purporting to teach people how to draw by exploiting the spatial and creative powers of the right brain.[*] This book outsold even Ornstein's book and remains a best-seller. In his 1977 book *The Dragons of Eden*, the noted cosmologist and populariser of science Carl Sagan portrayed the right hemisphere as the creative but paranoid instigator of scientific ideas, often seeing patterns and conspiracies where they do not exist. The role of the rational left hemisphere is to submit these ideas to critical scrutiny.

---

[*] I do not mean to disparage Edwards' teaching techniques, which many have no doubt found to be effective. Her reference to the right brain, though, may be largely superfluous.

The right brain insinuated its way into the business world. In 1976, a professor in the Faculty of Management at McGill University was moved to write in the *Harvard Business Review* as follows:

> The important policy processes of managing an organization rely to a considerable extent on the faculties identified with the brain's right hemisphere. Effective managers seem to revel in ambiguity; in complex, mysterious systems with no order.

This must have had an impact. Google 'right-brain business' today and you'll find around 350 million entries.

More critical analysis suggests that all may not be right. In one recent study, students of art and design were asked to create book cover illustrations while their brain activity was monitored in an MRI scanner. Although these students were artistically inclined and were engaged in an art project, there was no evidence that they were drawing on the right sides of their brains. Instead, the areas activated included regions in the frontal lobes associated with executive functions, along with the default-mode network under-lying mind-wandering. Neither side of the brain was favoured.

In a more extensive review of evidence from brain-imaging associated with measures of creative cognition, Rex Jung and his colleagues conclude that, as a 'first approximation', creativity depends on the very mechanisms of mind-wandering itself—the default-mode network. It is surely more likely that the source of creativity is to be found in widespread networks in the brain, rather than cramped up on the right side—even in the brain, the further we wander, the more likely we are to find something new: a new connection, as Steve Jobs put it. Edward de Bono, sometimes called the father of creativity, encourages his audiences to 'think outside the box'. He does not endorse the right-brain theory, although he does add an interesting twist: 'We believe the right side of the brain represents creativity, but it does not. It represents innocence, which may play a role in creativity—particularly in artistic expression.' There may well be a morsel of truth in the idea that the right brain is the more involved in artistic creativity, the left in verbal creativity, but we should relinquish our obsession with brain duality, and let the whole brain get on with it.

If creativity depends on widespread networks, you might expect more long-range connections in creative individuals than in the non-creative. Such connections make up the white matter of the brain, and one study showed divergent thinking to be unrelated to white-matter volume in either the left or the right brain. Rather surprisingly, though, more creative individuals tended to have smaller corpora callosa. The authors of this study suggested that a smaller corpus callosum provides greater independence in each side of the brain. Perhaps creativity depends not so much on thinking outside the box as on having two boxes to think with. Kipling may have been right.

On randomness

The distinguished psychologist and epistemologist Donald T. Campbell (1916–1996) once described the essence of creativity as 'blind variation and selective retention'. Blind variation is captured in the very notion of wandering, whether ambulatory or mental—straying from a set path into unknown territory. What we find there then depends on chance. It is the randomness of our wanderings, then, that supplies the spark of creativity, although when we do stumble across something new and important we need to recognise it as such—what Campbell called 'selective retention'.

Indeed, randomness seems to permeate the very universe, and not just our fickle minds. According to the uncertainty principle in physics, even subatomic particles don't seem to know exactly where they are. Or more accurately, you can't specify precisely both the location and momentum of a particle—the more you know of one, the less you know of the other. They can therefore only be located according to a probability distribution, as though wandering in their rather confined niches, fighting for their own space. Albert Einstein famously remarked to Max Born that 'God does not play dice with the universe', but perhaps that's precisely what God, if he or she exists outside of a probability distribution, actually does.

We will also never know exactly what the weather will do next, where each raindrop will fall, or where and when the next earthquake will occur—in New Zealand, the devastating earthquakes in Christchurch in 2010 and 2011 came as almost complete surprises, despite the diligent research of seismologists. The emergence of life itself on this lucky planet was also a matter of happenstance—the right mix of primordial soup and maybe a lightning strike to set the thing going. Once it started, randomness played the major role in building the pulsating planet we inhabit today. Evolution capitalises on random changes to the genome that add to survival value, and we are ourselves the product of a vast number of random events, selected and ratcheted over time. Learning, too, seems to depend on random activity. Even the behaviourists understood that behaviour must be 'emitted' before it can be reinforced, and stamped into an animal's repertoire. The pigeon that insistently pecks a key to receive a food reward must first peck it by chance, and only then discover the benefits of doing so.

Moving animals are disposed to wander through space. Sometimes they do so in goal-directed fashion, taking the well-trodden route to the watering hole, or the traffic-snarled road to work. But sometimes they just roam, exploring new territories, or perhaps wondering as they wander what's around the next bend. This too can lead to evolutionary change. Migration patterns in birds may have begun because birds that wandered found better conditions, and produced more offspring, than the stay-at-homes. For instance, birds originating in the tropics may have discovered that by wandering north they could enjoy longer daylight hours, allowing them to raise more young. They would then return before the northern winter, as the days shortened. Such patterns were eventually incorporated into the genetic make-up. We humans are prolific wanderers, having dispersed from Africa around 70,000 or so years ago to populate the globe. Except for the Canadians who migrate to Florida and New Zealanders to the Australian Gold Coast in winter, human wandering is for the most part exploratory, leading to the discovery of new lands, new climates, new ways of coping, greener grass.

It is through wandering, whether physical or mental, that we invite randomness to intervene, and so discover novelty. William Wordsworth found much of his poetic inspiration by wandering in the Lake District in the north-east of England:

I wandered lonely as a cloud
That floats on high o'er vales and hills,
When all at once I saw a crowd,
A host, of golden daffodils;
Beside the lake, beneath the trees,
Fluttering and dancing in the breeze.

His wandering was no doubt as much mental as physical, lending poetic voice to an experience that was itself a matter of chance.

The previous chapters in this book have covered some of the ways in which we wander in mind only, whether in mental time travel, imagining ourselves inhabiting the minds of others, dreaming, or hallucinating. All have a random element, taking us into mental territory that can prove unexpected and even illuminating. Much of our mental wandering, like spatial wandering, takes us into territories that are of no consequence for our futures— or those of our fellow humans. Occasionally, though, we strike gold.

Dreams are a form of uncontrolled mind-wandering that may lead to creative ideas, so long as we remember them. Otto Loewi won the Nobel Prize in Physiology or Medicine in 1936 for his work on the chemical transmission of nerve impulses, and is said to have discovered how to prove his theory in a dream. Robert Louis Stevenson developed the plot of *Dr Jekyll and Mr Hyde* in a dream. August Kekulé hit upon the ring shape of the benzene molecule after a daydream of a snake seizing its own tail—although some have doubted Kekulé's story. And Jack Nicklaus corrected his golf swing as a result of a dream.

But you shouldn't rely too much on dreams. William James told the story of a Mrs Amos Pinchot, who had a dream in which she believed she had discovered the secret of life. Half asleep, she quickly wrote it down. When fully awake, she saw what she had written:

Hogamus, Higamus
Man is polygamous
Higamus, Hogamus
Woman is monogamous.

Dreams aren't always as revelatory as they may have seemed at the time. And in any case, we forget nearly all of them.

The effects of mind-altering drugs may be a more potent source of inspiration, because they are manifest while we are awake and leave a more lasting impression. Like dreams, they are outside of our control, providing a strong dose of randomness, but they often make too little sense to lead to productive creativity. Nevertheless, many artists and writers have turned to drugs, sometimes with the explicit aim of finding inspiration and enlightenment. The English Romantic poets around the turn of the nineteenth century found much of their muse in opium, which began to be imported into England during the eighteenth century. Wordsworth had experimented with it, and it may well have added to the golden lustre of the daffodils. His friend Coleridge was much more reliant on opium for poetic inspiration. He started using opium to relieve his rheumatism, but then came to believe it harmonised his body with his soul, if not with Wordsworth's—the two friends fell out with one another as Coleridge's addiction grew. Two of his most famous epic poems, *The Rime of the Ancient Mariner* and *Kubla Khan*, are said to have resulted from opium-induced visions.

Thomas De Quincey started taking opium for an equally mundane reason, to alleviate toothache, but he too soon came to appreciate its power to transcend. In his 1821 work *Confessions of an English Opium-Eater,* he observed that 'happiness can now be bought for a penny', referring to the happy age when laudanum (a mixture of opium and alcohol) could easily be obtained cheaply from street vendors. Inflation and the illicit nature of the drug industry have since added to the price. De Quincey's descriptions of opium-induced dreams and altered consciousness influenced later writers, including Edgar Allan Poe, Charles Baudelaire and Nikolai Gogol, but he also told of the excruciating torment he suffered through his addiction.

A great many writers in the nineteenth century took opium in the search for inspiration, including Elizabeth Barrett Browning, Wilkie Collins, Charles Dickens, Arthur Conan Doyle, John Keats, Edgar Allan Poe, Sir Walter Scott, Percy Bysshe Shelley and Robert Louis Stevenson; one wonders where nineteenth-century literature would have been without it. It was not just writers. The American polymath, inventor and scientist Benjamin Franklin experimented with both hashish and opium. Twentieth-century users of opium include Billie Holiday, Jean Cocteau and Senator Joe McCarthy. Pablo Picasso said: 'The smell of opium is the least stupid smell in the world.'

Cannabis and its various products seem to have had a more lasting legacy, though perhaps as much for its recreational delights as for its revelatory powers. In the form of hashish, it is said to have been introduced into Europe by Napoleon's army, who discovered it in Egypt after a victory there. George Washington farmed cannabis plants, as did Thomas Jefferson. Also known as marijuana, it seems to have been a source of recreation for American politicians, including Thomas Jefferson, Al Gore, Bill Clinton, Newt Gingrich and US Supreme Court judge Clarence Thomas. Salvador Dalí said: 'Everyone should eat hashish, but only once.' He also said: 'I don't use drugs. I *am* drugs.'

A latecomer was LSD, first synthesised in 1938, and found to induce powerful hallucinations and distortions of thinking. Largely through the efforts of Timothy Leary at Harvard University, LSD became the drug of choice in the psychedelic 1960s. In his autobiography *Flashbacks*, Leary claimed that 75 per cent of the professors, students, graduate students, writers and professionals who took LSD trips found the experience to be the most educational and revealing of their lives. The English novelist and essayist Aldous Huxley also wrote in praise of drug-induced enlightenment, having experimented first with mescaline and then with LSD, and famously took 100 grams of LSD as he lay dying. His drug-induced experiences are described in his 1954 book of essays whose title *The Doors of Perception* was taken from a line in William Blake's book *The Marriage of Heaven and Hell*, written between 1790 and 1793. Blake's own writing and art had many of the features of drug-induced revelations, but there seems to be no evidence that he actually, in the 1960s vernacular, 'did drugs'. LSD also inspired musicians such as The Beatles, Jimi Hendrix, Jim Morrison, The Mothers of Invention and The Rolling Stones, as well as the actors Peter Fonda, Cary Grant and Jack Nicholson. Steve Jobs, co-founder of the Apple company, used both marijuana and LSD. Indeed, LSD might have helped create the computer industry as a whole, since Silicon Valley emerged in California at the same time as LSD exploded into the cultural scene.

And there's always been alcohol—perhaps the most widely sanctioned of mind-altering drugs, but in many respects the most dangerous. It was the drug of choice for Winston Churchill and Franklin D. Roosevelt, and lit the fires of creativity in many talented writers, including Truman Capote, John Cheever, Ernest Hemingway, William Faulkner, James Joyce, Jack Kerouac, Dorothy Parker and Dylan Thomas. In her novel *The Bell Jar*, Sylvia Plath wrote:[*] 'I began to think vodka was my drink at last. It didn't taste like anything, but it went straight down into my stomach like a sword swallower's sword and made me feel powerful and godlike.' Ogden Nash was more succinct: 'Candy is dandy but liquor is quicker.'

I am sure that people will continue to use drugs not only to find inspiration, but also simply for the transcendental experience. Drugs certainly add randomness to our thoughts, and in that sense can lead to creativity—more so, perhaps, in art and writing than in science. But there are, of course, serious downsides. One is that the induced randomness may be devoid of meaning—simply too much of a jumble to provide meaningful insight or aesthetic value. Another is that the sense of revelation itself turns out to be illusory in the cold light of sobriety. More serious, perhaps, is that many of the most potent drugs are addictive, and the pain of escaping their grip is simply too great to compensate for whatever pleasure of inspiration they brought in the first place. And besides, I feel bound to mention, most mind-altering drugs are illegal.

As we all know from attending a boring lecture, trying to listen to a symphony after a heavy meal, or simply trying to sleep on the

---

[*]   Originally published under the pseudonym 'Victoria Lucas'.

plane, the mind can wander quite freely without the influence of drugs. Even undirected wandering can stimulate creativity indirectly, through what is known as 'incubation', in which ideas are developed while one is thinking of something else. This has even been demonstrated experimentally. People were given the task of inventing unusual uses for familiar objects, a task commonly used as a measure of creativity. After working on this for a short time, most of them were given a break. During the break, some engaged in a task demanding of memory, some in an undemanding task, and some simply sat quietly without doing anything. When the creativity task resumed, those who performed the undemanding task performed best, probably because their minds wandered; other research has shown that undemanding tasks are most likely to induce mind-wandering, more likely even than doing nothing. If you're seeking inspiration, it seems a good idea to take a break and do something undemanding, like washing the dishes or watching a light TV show. Or perhaps knitting, which could explain why Agatha's Christie's Miss Marple, a compulsive knitter, was able to solve murder mysteries. Maybe Agatha Christie was herself a compulsive knitter, which is why she was able to create the murder mysteries in the first place.

An unnamed physicist is said to have told the German psychologist Wolfgang Köhler: 'We often talk about the three Bs, the Bus, the Bath, and the Bed. That is where the great discoveries are made in our science.'[*] He or she was perhaps alluding to Poincaré's mathematical inspiration that came to him as he stepped onto a bus, as well as to Archimedes' famous 'Eureka!' discovery that the

---

[*] Also attributed to the philosopher Ludwig Wittgenstein.

water rose when he was in the bath. As for the bed, dreams can sometimes lead to creative moments, but inspiration is perhaps more likely when we can't sleep, instead allowing our minds to wander while we're conscious enough to snare any insights. Perhaps one could add a fourth B, the boardroom, which provides an almost perfect environment for creative mind-wandering and incubation. And then there's boredom itself. The Nobel Prize-winning poet Joseph Brodsky once declared: 'Boredom is your window on the properties of time that one tends to ignore to the likely peril of one's mental equilibrium. It is your window on time's infinity. Once this window opens, don't try to shut it; on the contrary, throw it wide open.'

However you choose to wander, do not be discouraged into thinking that it is a waste of time. Of course, teacher was not always wrong—there are occasions when we need to attend in order to learn or finish some job. But nature also designed us to dream, to escape the channels that confine us. Remember from Chapter 1 the study by Jonathan Schooler and his associates on the frequency with which people zoned out while reading *War and Peace*? Well, those whose minds wandered most scored best on various measures of creativity. If the teacher or the board chairman catches you looking out the window when important matters are under discussion, you can explain that you are simply opening the doors of creativity.

And if your mind occasionally wandered while reading this book, I hope it took you into territory that was stimulating, creative—and above all happy.

References

Chapter 1: Meandering Brain, Wandering Mind

Epel, E. S., Puterman, E., Lin, J., Blackburn, E., Lazaro, A. and Mendez, W. B. (2013). 'Wandering minds and aging cells'. *Clinical Psychological Science*, 1, 75–83.

Ingvar, D. H. (1979). '"Hyperfrontal" distribution of the cerebral grey matter flow in resting wakefulness: On the functional anatomy of the conscious state'. *Acta Neurologica Scandinavica*, 60, 12–25.

——. (1985). '"Memory of the future": An essay on the temporal organization of conscious awareness'. *Human Neurobiology*, 4, 127–136.

Killingsworth, M. A. and Gilbert, D. T. (2010). 'A wandering mind is an unhappy mind'. *Science*, 330, 932–932.

Ottaviani, C. and Couyoumdjian, A. (2013). 'Pros and cons of a wandering mind: A prospective study'. *Frontiers in Psychology*, 4, Article 524.

Raichle, M. E., MacLeod, A. M., Snyder, A. Z., Powers, W. J., Gusnard, D. A. and Shulman, G. L. (2001). 'A default mode of brain function'. *Proceedings of the National Academy of Sciences USA*, 98, 676–682.

Schooler, J. W., Reichle, E. D. and Halpern, D. V. (2005). 'Zoning-out during reading: Evidence for dissociations between experience and meta-consciousness'. In D. T. Levin (ed.), *Thinking and Seeing: Visual Metacognition in Adults and Children* (pp. 204–226). Cambridge, MA: MIT Press.

Subramaniam, K. and Vinogradov, S. (2013). 'Improving the neural mechanisms of cognition through the pursuit of happiness'. *Frontiers in Human Neuroscience*, 7, Article 452.

Chapter 2: Memory

Corkin, S. (2002). 'What's new with the amnesic patient H.M.?'. *Nature Reviews Neuroscience*, 3, 453–460.

——. (2013). *Permanent Present Tense: The Man With No Memory, and What He Taught the World*. London: Allen Lane.

Kundera, M. (2002). *Ignorance*. New York: HarperCollins (translated from the
French by L. Asher).

Loftus, E. and Ketcham, K. (1994). *The Myth of Repressed Memory: False
Memories and Allegations of Sexual Abuse*. New York: St. Martin's Press.

Luria, A. R. (1968). *The Mind of a Mnemonist: A Little Book about a Vast Memory*.
London: Basic Books.

Martin, V. C., Schacter, D. L., Corballis, M. C. and Addis, D. R. (2011). 'A role for
the hippocampus in encoding simulations of future events'. *Proceedings of the
National Academy of Sciences USA*, 108, 13858–13863.

Nabokov, V. (2000). *Speak, Memory*. London: Penguin Books.

Ogden, J. A. (2012). *Trouble in Mind: Stories from a Neuropsychologist's Casebook*.
Oxford: Oxford University Press.

Raz, A., Packard, M. G., Alexander, G. M., Buhle, J. T., Zhu, H., Yu, S. and Peterson,
B. S. (2009). 'A slice of ϖ: An exploratory neuroimaging study of digit encoding
and retrieval in a superior memorist'. *Neurocase*, 15, 361–372.

Sacks, O. (1985). *The Man Who Mistook His Wife for a Hat and other Clinical
Tales*. New York: Simon & Schuster.

Spence, J. (1984). *The Memory Palace of Matteo Ricci*. London: Faber & Faber.

Tammet, D. (2009). *Embracing the Wide Sky*. New York: Free Press.

Treffert, D. A. and Christensen, D. D. (2006). 'Inside the mind of a savant'.
Scientific American Mind, 17, 55–55.

von Hippel, W. and Trivers, R. (2011). 'The evolution and psychology of self-
deception'. *Behavioral and Brain Sciences*, 34, 1–56.

Chapter 3: On Time

Clayton, N. S., Bussey, T. J. and Dickinson, A. (2003). 'Can animals recall the past
and plan for the future?'. *Trends in Cognitive Sciences*, 4, 685–691.

Darwin, C. (1896). *The Descent of Man, and Selection in Relation to Sex* (2nd
edition). New York: Appleton.

Markus, H. and Nurius, P. (1986). 'Possible selves'. *American Psychologist*, 41,
954–969.

Osvath, M. and Karvonen, E. (2012). 'Spontaneous innovation for future
deception in a male chimpanzee'. *PLOS ONE*, 7, e36782.

Suddendorf, T. and Corballis, M. C. (2007). 'The evolution of foresight: What
is mental time travel, and is it unique to humans?'. *Behavioral and Brain
Sciences*, 30, 299–351.

Suddendorf, T. and Redshaw, J. (2013). 'The development of mental scenario building and episodic foresight'. *Annals of the New York Academy of Sciences,* 1296, 135–153.

Tulving, E. (1985). 'Memory and consciousness'. *Canadian Psychologist,* 26, 1–12.

Wearing, D. (2005). *Forever Today: A Memoir of Love and Amnesia.* New York: Doubleday.

Chapter 4: The Hippo in the Brain

Addis, D. R., Wong, A. T. and Schacter D. L. (2007). 'Remembering the past and imagining the future: Common and distinct neural substrates during event construction and elaboration'. *Neuropsychologia,* 45, 1363–1377.

Corballis, M. C. (2013). 'Mental time travel: The case for evolutionary continuity'. *Trends in Cognitive Sciences,* 17, 5–6.

Dalla Barba, G. and La Corte, V. (2013). 'The hippocampus, a time machine that makes errors'. *Trends in Cognitive Sciences,* 17, 102–104.

Ekstrom, A. D., Kahana, M. J., Caplan, J. B., Fields, T. A., Isham, E. A., Newman, E. L. and Fried, I. (2003). 'Cellular networks underlying human spatial navigation'. *Nature,* 425, 184–187.

Gross, C. G. (1993). 'Huxley versus Owen: The hippocampus minor and evolution'. *Trends in Neurosciences,* 16, 493–498.

Macphail, E. M. (2002). 'The role of the avian hippocampus in spatial memory'. *Psicologica,* 23, 93–108.

Maguire, E. A., Woollett, K. and Spiers, H. J. (2006). 'London taxi drivers and bus drivers: A structural MRI and neuropsychological analysis'. *Hippocampus,* 16, 1091–1101.

Milivojevic, B. and Doeller, C. F. (2013). 'Mnemonic networks in the hippocampal formation: From spatial maps to temporal and conceptual codes'. *Journal of Experimental Psychology: General.* Advance online publication. doi: 10.1037/a0033746.

O'Keefe, J. and Nadel, L. (1978). *The Hippocampus as a Cognitive Map.* Oxford: Clarendon Press.

Pastalkova, E., Itskov, V., Amarasingham, A. and Buzsáki, G. (2008). 'Internally generated cell assembly sequences in the rat hippocampus'. *Science,* 321, 1322–1327.

Smith, D. M. and Mizumori, S. J. Y. (2006). 'Hippocampal place cells, context, and episodic memory'. *Hippocampus,* 16, 716–729.

Suddendorf, T. (2013). 'Mental time travel: continuities and discontinuities'. *Trends in Cognitive Sciences*, 17, 151–152.

Chapter 5: Wandering into other Minds

Bloom, P. (2004). *Descartes' Baby: How the Science of Child Development Explains What Makes Us Human*. New York: Basic Books.

Call, J. and Tomasello, M. (2008). 'Does the chimpanzee have a theory of mind? 30 years later'. *Trends in Cognitive Sciences*, 12, 187–192.

Darwin, C. (1872). *The Expression of the Emotions in Man and Animals*. London: John Murray.

de Waal, F. B. M. (2012). 'The antiquity of empathy'. *Science*, 336, 874–876.

Grandin, T. and Johnson, C. (2005). *Animals in Translation: Using the Mysteries of Autism to Decode Animal Behavior*. New York: Scribner.

Hare, B. and Woods, V. (2013). *The Genius of Dogs: How Dogs are Smarter than You Think*. London: Oneworld Publications.

Humphrey, N. (1976). 'The social function of intellect'. In P. P. G. Bateson and R. A. Hinde (eds), *Growing Points in Ethology* (pp. 303–317). Cambridge, UK: Cambridge University Press.

Kovács, A. M., Téglás, E. and Endress, A. D. (2011). 'The social sense: Susceptibility to others' beliefs in human infants and adults'. *Science*, 330, 1830–1834.

Laing, R. D. (1970). *Knots*. London: Penguin.

Marks, D. F. and Kammann, R. (1980). *The Psychology of the Psychic*. Buffalo, NY: Prometheus Books.

Premack, D. and Woodruff, G. (1978). 'Does the chimpanzee have a theory of mind?'. *Behavioral and Brain Sciences*, 1, 515–526.

Radin, D. I. (2006). *Entangled Minds: Extrasensory Experiences in a Quantum Reality*. New York: Paraview Pocket Books.

Randi, J. (1982). *The Truth About Uri Geller*. New York: Prometheus Books.

Suddendorf, T. (2013). *The Gap: The Science of What Separates Us from Other Animals*. New York: Basic Books.

Suddendorf, T. and Corballis, M. C. (1997). 'Mental time travel and the evolution of the human mind'. *Genetic, Social, and General Psychology Monographs*, 123, 133–167.

Taylor, M. (1999). *Imaginary Companions and the Children Who Create Them*. New York: Oxford University Press.

Whiten, A. and Byrne, R. W. (1988). 'Tactical deception in primates'. *Behavioral and Brain Sciences*, 11, 233–273.

Wimmer, H. and Perner, J. (1983). 'Beliefs about beliefs: Representation and constraining function of wrong beliefs in young children's understanding of deception'. *Cognition*, 13, 103–128.

Chapter 6: Stories

Abrahams, R. D. (1970). *Deep Down in the Jungle: Negro Narrative Folklore from the Streets of Philadelphia*. Chicago: Aldine.

Bateson, G. (1982). 'Difference, double description and the interactive designation of self'. In F. Allan Hanson (ed.), *Studies in Symbolic and Cultural Communication* (pp. 3–8). University of Kansas Publications in Anthropology No. 14. Lawrence: University of Kansas Press.

Boyd, B. (2009). *On the Origin of Stories: Evolution, Cognition, and Fiction*. Cambridge, MA: Belknap Press of Harvard University Press.

Corballis, M. C. (2002). *From Hand to Mouth: The Origins of Language*. Princeton: Princeton University Press.

Dunbar, R. I. M. (1998). *Grooming, Gossip, and the Evolution of Language*. Cambridge, MA: Harvard University Press.

Engel, S. (1995). *The Stories Children Tell: Making Sense of the Narratives of Childhood*. New York: W. H. Freeman.

Janet, P. (1928). *L'Évolution de la mémoire et de la notion du temps: Leçons au Collège de France 1927–1928*. Paris: Chahine.

Kidd, D. C. and Castano, E. (2013). 'Reading literary fiction improves theory of mind'. *Science*, 342, 377–380.

Mechling, J. (1988). '"Banana cannon" and other folk traditions between humans and nonhuman animals'. *Western Folklore*, 48, 312–323.

Niles, J. D. (2010). *Homo Narrans: The Poetics and Anthropology of Oral Literature*. Philadelphia: University of Pennsylvania Press.

Salmond, A. (1975). 'Mana makes the man: A look at Maori oratory and politics'. In M. Bloch (ed.), *Political Language and Oratory in Traditional Society* (pp. 45–63). New York: Academic Press.

Savage-Rumbaugh, S., Shanker, S. G. and Taylor, T. J. (1998). *Apes, Language, and the Human Mind*. New York: Oxford University Press.

Sugiyama, M. S. (2011). 'The forager oral tradition and the evolution of prolonged juvenility'. *Frontiers in Psychology*, 2, Article 133.

Thompson, T. (2011). 'The ape that captured time: Folklore, narrative, and the human-animal divide'. *Western Folklore*, 69, 395–420.

Trinkaus, E. (2011). 'Late Pleistocene adult mortality patterns and modern human establishment'. *Proceedings of the National Academy of Sciences USA*, 108, 1267–1271.

Turton, D. (1975). 'The relationship between oratory and the emergence of influence among the Mursi'. In M. Bloch (ed.), *Political Language and Oratory in Traditional Society*. New York: Academic Press.

Chapter 7: Tigers in the Night

Darwin, C. (1872). *The Expression of the Emotions in Man and Animals*. London: John Murray.

Foulkes, D. (1999). *Children's Dreaming and the Development of Consciousness*. Cambridge, MA: Harvard University Press.

Fox, K. C. R., Nijeboer, S., Solomonova, E., Domhoff, G. W. and Christoff, K. (2013). 'Dreaming as mind wandering: Evidence from functional neuroimaging and first-person content reports'. *Frontiers in Psychology*, 7, Article 412.

Freud, S. (1900). *The Interpretation of Dreams*. New York: Macmillan.

Hobson, J. A. (2009). 'REM sleep and dreaming: Towards a theory of protoconsciousness'. *Nature Reviews Neuroscience*, 10, 803–813.

Honikawa, T., Tamaki, M., Miyawaki, Y. and Kamitani, Y. (2013). 'Neural decoding of visual imagery during sleep'. *Science*, 340, 630–642.

Revonsuo, A. (2000). 'The reinterpretation of dreams: An evolutionary hypothesis of the function of dreaming'. *Behavioral and Brain Sciences*, 23, 877–901.

Saurat, M.-T., Agbakou, M., Attigui, P., Golmard, J.-L. and Arnulf, I. (2011). 'Walking dreams in congenital and acquired paraplegia'. *Consciousness and Cognition*, 20, 1425–1432.

Valli, K. and Revonsuo, A. (2009). 'The threat simulation theory in light of recent empirical evidence: A review'. *American Journal of Psychology*, 122, 17–38.

Wamsley, E. J. and Stickgold, R. (2010). 'Dreaming and offline memory processing'. *Current Biology*, 20(23), R1010.

Chapter 8: Hallucinations

James, W. (1902). *The Varieties of Religious Experience: A Study in Human Nature*. London: Longmans, Green & Co.

Jaynes, J. (1976). *The Origin of Consciousness in the Breakdown of the Bicameral Mind*. New York: Houghton Mifflin.

McGilchrist, I. (2009). *The Master and his Emissary: The Divided Brain and the Making of the Western World*. New Haven, CT: Yale University Press.

Penfield, W. and Perot, P. (1963). 'The brain's record of auditory and visual experience'. *Brain*, 86, 596–696.

Rosenhan, D. L. (1973). 'On being sane in insane places'. *Science*, 179, 250–258.

Sacks, O. (2012). *Hallucinations*. New York: Random House.

Sireteanu, R., Oertel, V., Morh, H., Linden, D. and Singer, W. (2008). 'Graphical illustration and functional neuroimaging of visual hallucinations during prolonged blindfolding'. *Perception*, 37, 1805–1821.

Vitorovic, D. and Biller, J. (2013). 'Musical hallucinations and forgotten tunes—case report and brief literature review'. *Frontiers in Neurology*, 4, Article 109.

Waugh, E. (1957). *The Ordeal of Gilbert Pinfold*. London: Chapman & Hall.

Zubek, J. P. (ed.) (1969). *Sensory Deprivation: Fifteen Years of Research*. New York: Meredith.

Chapter 9: The Creativity of the Wandering Mind

Baird, B., Smallwood, J., Mrazek, M. D., Kam, J. W. Y., Franklin, M. S. and Schooler, J. W. (2012). 'Inspired by distraction: Mind wandering facilitates creative incubation'. *Psychological Science*, 23, 1117–1122.

Campbell, D. T. (1960). 'Blind variation and selective retention in creative thought as in other knowledge processes'. *Psychological Review*, 67, 380–400.

Corballis, M. C. (1999). 'Are we in our right minds?'. In S. Della Sala (ed.), *Mind Myths: Exploring Popular Assumptions About the Mind and Brain* (pp. 25–42). Chichester: John Wiley & Sons.

de Bono, E. (1995). 'Serious creativity'. *The Journal for Quality and Participation*, 18, 12–19.

De Quincey, T. (1822). *Confessions of an English Opium-Eater*. London: Taylor & Hessey.

Edwards, B. (1979). *Drawing on the Right Side of the Brain*. New York: Penguin Putnam.

Ellamil, M., Dobson, C., Beeman, M. and Christoff, K. (2012). 'Evaluative and

generative modes of thought during the creative process'. *NeuroImage*, 59,
1783–1794.

Huxley, A. (1954). *The Doors of Perception*. London: Chatto & Windus.

Jung, R. E., Mead, B. S., Carrasco, J. and Flores, R. E. (2013). 'The structure of
creative cognition in the human brain'. *Frontiers in Human Neuroscience*, 7,
Article 300.

Leary, T. (1983). *Flashbacks: A Personal and Cultural History of an Era*.
Los Angeles: Tarcher.

Lucas, V. [Plath, S.] (1963). *The Bell Jar*. London: Heinemann.

Moore, D. W., Bhadelia, R. A., Billings, R. L., Fulwiler, C., Heilman, K. M., Rood,
K. M. J. and Gansler, D. A. (2009). 'Hemispheric connectivity and the visual-
spatial divergent-thinking component of creativity'. *Brain and Cognition*, 70,
267–272.

Mintzberg, H. (1976). 'Planning on the left side and managing on the right'.
Harvard Business Review, 54, 49–58.

Ornstein, R. E. (1972). *The Psychology of Consciousness*. New York: Harcourt
Brace.

Sagan, C. (1977). *The Dragons of Eden: Speculations on the Evolution of Human
Intelligence*. New York: Random House.

Sperry. R.W. (1982). 'Some effects of disconnecting the cerebral hemispheres'.
Science, 217, 1223-1226.

Stevenson, R. L. (1886). *Strange Case of Dr Jekyll and Mr Hyde*. London:
Longmans, Green & Co.

Index

Page numbers in *italic* denote references.